BUILDING MAINTENANCE MANAGEMENT

Second Edition

Barrie Chanter
BSc(Hons) MSc MCIOB

and

Peter Swallow
Dip. Surv. Dip. Arch. Cons. FRICS FBEng. FRSA

Blackwell
Publishing

Blackwell Publishing editorial offices:
Blackwell Publishing Ltd, 9600 Garsington Road, Oxford OX4 2DQ, UK
Tel: +44 (0)1865 776868
Blackwell Publishing Inc., 350 Main Street, Malden, MA 02148-5020, USA
Tel: +1 781 388 8250
Blackwell Publishing Asia Pty Ltd, 550 Swanston Street, Carlton, Victoria 3053, Australia
Tel: +61 (0)3 8359 1011

First published 1996 by Blackwell Publishing Ltd
Second edition published 2007

ISBN: 978-1-4051-3506-1

Library of Congress Cataloging-in-Publication Data

Chanter, Barrie.
Building maintenance management / Barrie Chanter and Peter Swallow. – 2nd ed.
p.   cm.
Includes bibliographical references and index.
ISBN-13: 978-1-4051-3506-1 (pbk. : alk. paper)
1. Buildings–Maintenance–Management.   I. Swallow, Peter.   II. Title.

TH3361.C47 2006
658.2′02–dc22

2006037651

A catalogue record for this title is available from the British Library

Set in 10 on 12.5 pt Palatino
by SNP Best-set Typesetter Ltd., Hong Kong

For further information on Blackwell Publishing, visit our website:
www.blackwellpublishing.com/construction

BUILDING MAINTENANCE MANAGEMENT

Contents

Preface to the Second Edition

The initial impetus for the second edition of this text was to carry out a general review and update, along with the introduction of a new chapter to address facilities management. At the time of writing the first edition, facilities management was an emerging issue clearly impinging on the practice of maintenance management. Since that time its growth has been rapid.

So too, however, has been the rate of change of other important contextual issues. This persuaded us to broaden the scope of the second edition, and the new Chapter 1 goes beyond the original intentions, so as to more comprehensively embrace this changing context.

As the reader will appreciate, there are difficulties in this as the context continues to change. We have attempted to address the major contemporary developments in Chapter 1 initially, with consequential revisions of later chapters, notably in relation to maintenance organisations and execution of maintenance.

Other chapters have, we hope, been suitably refreshed and updated in line with current legislation, contractual forms, statistical data and government policy/guidelines.

As we comment in the footnote in Chapter 1, a major problem in compiling the text has been an environment of change that directly affects not only maintenance work but also sources of guidance and data. In particular, governmental re-shuffles of departments and personnel are troublesome when quoting sources. In general we have made the decision to retain the original references as at the time of their sourcing. We apologise for any confusion this may cause the reader, but felt it was the neatest option.

Barrie Chanter
Peter Swallow
January 2007

Preface to the First Edition

The growth in the significance of building maintenance as a proportion of the output of the construction industry has taken place against a backdrop of mounting pressure on new-build activity, and a growing awareness of the need to manage the condition of the nation's building stock more effectively. Paralleling these developments has been the increased application of new technology, permitting the more efficient use of data. Notwithstanding this, it is still the case that much maintenance activity takes place in a context that does not create a fully integrated approach to managing building performance and, thus, the full potential of many buildings is never wholly realised.

There is a wealth of information and informed comment available concerning the poor condition of much of the UK's building stock and this formed a natural starting point for Chapter 1. This reviews some of the currently available information, supplemented by an overview of the nature and significance of the repair and maintenance sector, within the context both of the construction economy and the national economy as a whole. The position of the maintenance department within the structure of the parent organisation is considered in Chapter 2 and this confirms the existence of a major weakness concerning clients' perceptions of the importance of maintaining the building fabric. We are indebted here to several individuals and organisations for providing valuable insights, from the perspective of the professional maintenance practitioner, which enabled us to analyse a range of typical organisational structures, both as exemplars and as an indication of the diversity of the thinking that exists. Our grateful thanks go to Stafford Taylor and his colleagues at Building Surveying Associates, together with all those others who, for reasons of confidentiality, it would not be appropriate to name here.

That poor detailed design affects building performance, and hence maintenance, is well known. Chapter 3 explores the design/maintenance relationship against the broader backcloth of the building procurement process. Many of the problems encountered in buildings stem from the brief development phase, where a failure to establish user requirements in sufficient detail results in the poor performance of the completed building. At the hand-over stage also there may be serious shortcomings, and more careful consideration must be given to providing the client with a proper building model to facilitate the effective management of the property. None of these crucial developments can take place without a major shift in client attitudes, and professions working with the built environment must shoulder some responsibility for this in their failure to educate clients sufficiently. For too long they have evaded responsibility for adequately preparing uninformed clients, and the increasingly competitive pressures of competitive fee tendering are a serious obstacle in this respect.

The nature of maintenance work is examined in some detail in Chapter 4 because a full understanding of the types of maintenance work is necessary in order to appreciate the manner in which the maintenance workload is generated. However, this has to be tempered with a degree of realism and a recognition of the context within which it occurs.

A large quantity of data is generated by maintenance operations, and its management is a complex matter. Chapter 5, therefore, focuses specifically on information management and develops the notion of a building condition model. The technology to produce and maintain an electronic performance model now exists at a variety of levels of sophistication. A number of software packages are described to provide a flavour of current practice. We would like to thank many people for their help in this area. Steve Wilson of Minster General Housing Association; the staff of Building Surveying Associates; colleagues at De Montfort University, notably Donna Wilson, Tracey Burt and Tony Gibbs in Property Services, Stuart Planner in Landstaff, Chris Watts of the Building Surveying Department, who provided invaluable guidance throughout the preparation of this chapter, and Rob Ashton of the CAD Centre who acted as an excellent specialist IT advisor. In developing the chapter on information management we were aware of two major factors. Firstly, the pace of software development is now so rapid that it is almost impossible to keep abreast of it, and thus it is only possible to provide a limited overview. Secondly, it was also apparent that in separating out maintenance planning in Chapter 6, there was some risk of duplication. However, it was felt that maintenance planning warranted separate consideration in order to clearly define good principles. Chapter 5 and 6 therefore should be read together.

Chapter 7, which deals with maintenance contracts, logically precedes the consideration of maintenance execution, the subject of Chapter 8. Inevitably there is some overlap from our consideration of detailed execution into a number of other areas, which we recognise but justify on the grounds that it is the context that is important.

Overlying the whole subject of maintenance management is the issue of facilities management. Throughout the text we have made reference to this rapidly developing discipline, particularly with respect to information management and the construction of buildings, whether they are viewed as an investment, asset or facility in the widest sense. Perhaps the biggest service provided by the growth of facilities management has been to focus attention more sharply on, and to promote the profile and image of, the people who manage buildings.

It is hoped that this book will both act as an update for practitioners and provide an introduction for final year students, graduates and others encountering building maintenance for the first time.

In presenting this introduction we have thanked a number of people who have provided material or acted as sounding boards for some of our ideas, but the biggest thank you of all must go to our respective wives, Lesley and Lynne, to whom we dedicate this book.

Barrie Chanter
Peter Swallow

About the Authors

Barrie Chanter is a chartered builder with a background in civil engineering and project management. He joined the School of Architecture at De Montfort University, Leicester, in September 1984 and moved to the Department of Building Surveying in 1988, where he led the undergraduate course in Building Surveying for ten years. With the formation of the Faculty of Art and Design, he continued to teach Professional Practice to architectural students but in 2002 was appointed to a Faculty management role to co-ordinate all collaborative projects at home and overseas. He retired from the university in July 2006 but continues to act as an advisor for a number of educational institutions with particular reference to the development of foundation degrees and other collaborative projects between the HE and FE sectors.

Professor Peter Swallow is a chartered building surveyor who joined the full-time staff at De Montfort University, Leicester, in 1974, following a varied career in both public and private sectors. He ran the Department of Building Surveying until the reorganisation of the University which led to the formation of the Faculty of Art and Design, when he became a member of the Leicester School of Architecture. He co-ordinates the teaching of professional practice throughout the school, up to RIBA final part three. He has lectured widely to professional audiences on building defects, repair and conservation issues. He is the author of *Measurement and Recording of Historic Buildings* (second edition, with Sophie Jackson, Jonathan Godfrey and Ross W.A. Dallas, 2004).

Acknowledgements

In producing the second edition of this text we have been grateful for help from a number of key sources. We would in particular like to thank Intelligent Systems Ltd and FMx Ltd for information on their software, allowing us to refresh Chapter 6.

We also owe a debt for the input of a number of part-time Building Surveying students, who over the years have given us invaluable input from the practitioners' perspective. In particular we would like to single out Phil Derbyshire, Dave Massingham and John Pallett for the intelligence they provided on key areas in the local government and health sectors.

As before, we would like to thank our wives Lynne and Lesley for their forbearance and, as with the first edition, dedicate the text to them.

Abbreviations

ACE	Association of Consulting Engineers
ACOP	Approved Codes of Practice
ADB	Activity DataBase
ADC	Association of District Councils
ALMO	Arm's Length Management Organisation
AMP	Asset Management Plan
BEC	Building Employers Confederation
BIFM	British Institute of Facilities Management
BMCIS	Building Maintenance Cost Information Service
BMI	Building Maintenance Information Ltd
BVPP	Best Value Performance Plan
BVPI	Best Value Performance Indicators
CA	contract administrator
CAD	computer-aided draughting
CCPI	Co-ordinating Committee for Project Information
CCT	Compulsory Competitive Tendering
CDM	Construction (Design & Management) Regulation
CIC	Construction Industry Council
CIOB	Chartered Institute of Building
CITB	Construction Industry Training Board
CLASP	Consortium of Local Authorities Special Programme
CPA	Comprehensive Performance Assessment
CPIC	Construction Project Management Information Committee
DBFO	Design Build Finance Operate
DES	Department for Education and Science
DFEE	Department for Education and Employment
DfES	Department for Education and Skills
DLO	Direct Labour Organisation
DoE	Department of the Environment
DIY	do-it-yourself
DQI	Design Quality Indicator
DTI	Department of Trade and Industry

EHCS	English House Condition Survey
ERIC	Estate Return Information Collection
FM	facilities management
GDP	gross domestic product
GNP	gross national product
HASWA	Health and Safety at Work etc. Act
HEC	Higher Education Corporation
HRA	Housing Revenue Account
HSC	Health and Safety Commission
HSE	Health and Safety Executive
IC	Intermediate Building Contract
IT	information technology
JCT	Joint Contracts Tribunal
KPI	key performance indicator
LCC	life cycle cost
LEA	local education authority
LMS	Local Management of Schools
MHSWR	Management of Health & Safety at Work Regulations
MRA	Major Repairs Allowance
MTC	Measured Term Contract
MW	Minor Works Building Contract
NBS	National Building Specification
NDS	New Deal for Schools
NEDO	National Economic Development Office
NFHA	National Federation of Housing Associations
NHS	National Health Service
NPV	net present value
ODPM	Office of the Deputy Prime Minister
OMR	optical mark reader
PC	prime cost
PFI	Private Finance Initiative
POE	Post-Occupancy Evaluation
PPE	personal protective equipment
PPP	Public Private Partnership
PSA	Property Services Agency
PUWER	Provision and Use of Work Equipment Regulations
RIBA	Royal Institute of British Architects
RICS	Royal Institution of Chartered Surveyors
RPI	retail price index
RRFSO	Regulatory Reform (Fire Safety) Order
RSL	Registered Social Landlord
SBC	Standard Building Contract
SCALA	Society of Chief Architects in Local Authorities
SO	supervising officer

SSP	Strategic Service Delivery Partnership
TFM	total facilities management
TSO	The Stationery Office
VAT	value added tax
WHSW	Workplace (Health, Safety and Welfare) Regulations

Chapter 1

The Changing Context within which Building Maintenance Operates

Over recent years there have been many developments of a generic nature that have had a profound effect on the manner in which estates are managed and maintained. A significant driver has been the rapid growth of facilities management, and connected with this, directly and indirectly, have been initiatives in procurement strategies, contracting out, and performance monitoring and measurement. Whilst these changes have affected both private and public sectors, there have been particularly important changes in the latter, which have had important repercussions for the context in which the maintenance of their building stock is carried out.

This chapter examines a number of these developments but uses as a starting point the growth of facilities management (FM). Whilst many of the issues discussed have a relationship with FM, it is emphasised that in some cases they are of wider significance in relation to managing building maintenance.

Facilities management

For some time it has been clear that managing buildings or estates has been carried out in the context of what has become known as facilities management. The scope of FM has now increased beyond early concepts and has taken a massive hold of strategic thinking in the management of buildings to the point where any sensible study of maintenance management cannot take place without reference to it. Associated with it are a number of concepts with which it has become synonymous such as outsourcing, service level agreements, ideas of best practice or value, and new approaches in procurement where a more holistic view of building life is implied.

The growth of FM as a key service industry has been extremely rapid and there have been a number of key reports analysing its size and composition. A difficulty in this respect is one of definition. Of some interest is the question of whether or not one should consider it as a discipline, profession or a strategic management concept. For our purposes we have preferred to keep to a rather more conceptual treatment.

Various definitions have emerged, including the following:

The infrastructure that supports the people in the organisation in their endeavors to achieve business goals.

The practice of co-ordinating the physical workplace with the people and work of the organisation; integrates the principles of business information, architecture and the behavioral and engineering sciences.

(USA Library of Congress)

Facilities management is a distinct management function and, as such, involves a well defined and consistent set of responsibilities. Simply stated, it is management of a vital asset – the organization's facilities . . . Facility management combines proven management with current technical knowledge to provide humane and effective work environments. It is the business practice of planning, providing and managing productive work environments.

(International Facility Management Association)

Facilities management is the integration of multi-disciplinary activities within the built environment and the management of their impact upon people and the workplace. Effective facilities management is vital to the success of an organisation by contributing to the delivery of its strategic and operational objectives.

(British Institute of Facilities Management)

It should be noted that these definitions do not necessarily, within themselves, reflect the range of services offered by FM service providers, which may range from the sophisticated to the apparently mundane. They do need, however, to be considered holistically as they all contribute to the proper (and efficient) management of a built asset/facility.

Broadly, FM divides itself into three operational facets:

❑   in-house management of facilities
❑   contracted out individual packages/contracts/services – e.g. catering
❑   total facilities management (TFM), sometimes referred to as strategic FM, where an integrated service is 'bought in' by the organisation or, conversely, contracted out.

To these it may be argued we should add Public Private Partnerships (PPP) and in particular the Private Finance Initiative (PFI), which, although much broader conceptually, is of massive significance and merits specific attention later.

Estimations of the market value for facilities management activity are difficult to arrive at, particularly when one also includes manpower, training and service level management costs. However, the value is huge, leading Roland Gribbens[1] to comment that facilities management is the boom industry of the twenty-first century.

A report for the Facilities Management Faculty of the RICS in September 2003[2] estimated the UK market for FM to be £94.9 billion in 2002, an increase of 35% in the period from 1998. For integrated FM, MSI[3] placed the value of the market at £4.5 billion in 2001 and reported that the largest portion was accounted for by the public sector, circa 57%. This is seen to be as a result of increasing attention being paid to a holistic approach rather than the creation of complex patterns of agreement through a multi-plicity of individual contracts. This thinking is in turn driving a move towards the formation of strategic partnerships and a significant move into this market by major contractors. This is fuelled by such contractors who seek to broaden their markets, raise their image and create consistent workloads. In a number of cases contractors have absorbed existing FM service providers to enhance the diversity of skills they possess and give them competitive advantage.

The RICS report also estimated that in-house FM had reduced its share of the market to 36%, and predicted that the TFM or Integrated FM market will increase to £10.4 billion up to 2007 and the market value as a whole will exceed £100 billion.

The move towards contracting out or 'outsourcing' is presumed to be client driven as organisations seek to focus on core activity and divest themselves of the responsibil-ity for non-core activity. Thus the major activities to be outsourced include catering, cleaning and security, to which we can increasingly add maintenance.

The origins and reasons for this growth are complex and it has given rise to a number of important associated issues. The origins of FM are probably in North America, but its spread to Europe and the Far East has been rapid and is a result of both supply and demand factors. Professional service providers and new entrants seek new markets whilst business organisations and public sector bodies have re-appraised the nature, characteristics and role of their properties and the manner in which they are managed. Although FM has broadened itself to reach beyond property or building management, for the purposes of this text we will confine ourselves to buildings and their management.

In this view of FM we can identify a clear and significant change in emphasis away from the traditional view of property as an investment, towards consideration of build-ings as a factor of production. As a consequence, maintenance management has become more firmly positioned in the context of strategic management in both public and private sectors. Indeed, as pointed out above, strategic FM in the public sector has been a major growth area.

A survey of a range of reports conducted by McGregor and Then identified a number of conclusions with respect to the conventional approach to property management:

❑ Reactive rather than proactive property management . . . property only seriously considered by organisations when they were under severe profit or cost constraints.
❑ Only on rare occasions does property receive explicit treatment in corporate plans.
❑ More often than not property is viewed as incidental, as an asset that requires little management, generates cost but has little or no value.
❑ . . . the whole area of monitoring organisational property assets is new . . .
❑ . . . few (companies surveyed) had a property strategy for their operations that amounted to more than 'we'll find space when we need it'.
❑ . . . property seen as a cost of business rather than a business resource . . .

❏   . . . real estate lags behind other key resource areas in terms of attention given and performance achieved.
❏   Given the acknowledged link between the workplace environment, employee satisfaction and profitability, senior managers do appear to be missing an opportunity to manage the working environment for competitive advantage.
❏   Space planning must play a bigger part in overall business development . . .

From this one could identify a scenario where too often we may have:

❏   corporate management that ignores the role of property or considers it to be a liability
❏   management viewing property purely in investment terms with a consequent over-reliance on the property market to 'bail them out'
❏   property departments operating in isolation, unaware of broad corporate objectives
❏   lack of information on the property portfolio which may be only an indication of a much more deeply seated malaise
❏   conflict between property departments and client departments because of inconsistency of perspective and priority
❏   separation of property operations along professional lines – e.g. separation of designers from building management
❏   a non-strategic financial strategy that, for example, allocates resources separately for revenue and capital without considering the relationship between them, which may therefore give rise to higher cost in use
❏   a tendency to be over-preoccupied with physical elements and ignore location, function, space and value
❏   alternatively, a tendency to view commercial property in investment terms as judged by the market.

These perceived shortcomings underpin what can be seen as a 'push-pull' effect, described diagrammatically in figure 1.1, which as much as anything has driven the FM expansion and provided a basis for the strategic approach to property that typifies FM activity.

The scope of facilities management

In the early days practitioners of FM preoccupied themselves not only with issues of definition but also as to whether it represented a profession, discipline or simply a concept. The emergence of organised professional bodies such as the British Institute of Facilities Management (BIFM) have done much to address these concerns and, by establishing training and education programmes, given us a major insight into the scope as perceived by the professional.

An initial perhaps simplistic view suggests that the following list provides a view of the scope of FM:

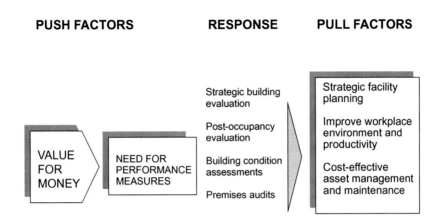

Figure 1.1 Push–pull factors.

❏ cleaning and waste disposal
❏ energy management
❏ environmental management
❏ estates management
❏ equipment and consumable purchasing
❏ fire safety
❏ grounds maintenance
❏ health and safety
❏ human resources
❏ office management
❏ property and engineering services maintenance
❏ relocation, refurbishment, adaptation, new build, etc.
❏ security
❏ space management and planning.

This list of activities, however, does little to reveal the nature of FM as it bypasses consideration of the management concepts that lie at its heart. There are a number of clues, given by the competencies identified by the BIFM and any number of FM provider internet sites that identify the scope of their activity.

The BIFM competencies[4] for their professional examinations are listed under the following headings:

❏ Understanding business organisation
❏ Managing people
❏ Managing premises
❏ Managing services
❏ Managing the working environment
❏ Managing resources.

The immediate concern of this text, managing building fabric, comes under the heading of Managing premises, and building services is one of a number of subheadings under Managing services.

Their Understanding facilities management (Foundation) course includes the following topics:

- ❏   the role of FM in achieving corporate success
- ❏   an introduction to buildings and services
- ❏   practical space planning and management
- ❏   re-housing the office
- ❏   energy management and environmental strategy
- ❏   maintenance management
- ❏   commissioning and managing contract services.

The Centre for Facilities Consultancy[5] gives the following as an indication of the scope of their services:

- ❏   strategic facilities planning
- ❏   staffing structure
- ❏   accommodation strategy
- ❏   customer surveys
- ❏   space planning, office moves and relocation
- ❏   computerised facilities and maintenance systems
- ❏   output specifications
- ❏   outsourcing exercises
- ❏   performance measurement systems
- ❏   occupancy cost studies
- ❏   business continuity plans
- ❏   financial systems and cost control
- ❏   change management
- ❏   video conferencing facilities.

Considering the range of services now encompassed within FM, one is tempted to think it easier to ask the question as to what is not included in FM.

The answer, then, cannot really lie in what FM does but in the approach it brings, and it is at this level that its importance to maintenance management clearly articulates itself. The BIFM in their introduction to FM say:

Facilities Management is the integration of multi-disciplinary activities within the built environment and the management of their impact upon people and the workplace.

They go on to say:

Within this fast growing professional discipline, Facilities Managers have extensive responsibilities for providing, maintaining and developing myriad services. These range from property strategy, space management and communications infrastructure to building maintenance, administration and contract management.

Another way of approaching the question of scope is to look at the recipients of the service. Broadly we can distinguish between the private and public sectors. The initial impetus for the growth of FM was in the former, and within this context FM has responsibility to shareholders to preserve the value of the company assets and to increase profits through delivering optimum use from a built asset as a factor of production. In addition, a key aspect of an improved built asset management is an improved working environment leading to higher productivity, lower absenteeism and improved motivation. These improvements should in turn provide a better customer interface, with the attendant marketing benefits.

The public sector, although slightly slower to respond, is now paying increased attention to a more holistic manner of looking after its estates. This has been extended to strategies that consider more specifically the whole life of the building from procurement to disposal in a manner that embraces many of the principles of FM.

Synchronous with this, in both public and private sectors, has been the focus on core business and hence the search for the means for an organisation to divest itself of non-core activity. This is not without its share of controversy, particularly in the public sector, where this process is often seen as privatisation.

In both sectors, however, it is clear that a move towards the introduction of a more integrated facilities management service, whether by in-house means, outsourcing or a hybrid approach, represents a complex exercise in change management.

Implementation of facilities management services

The implementation of an FM service carries all the risks associated with any change in management strategy. To manage these risks it is convenient to think of implementation in terms of processes. A number of models exist that have a great deal of commonality in that they describe FM processes in terms of the following levels:

❑ strategic
❑ tactical
❑ operational.

In some other models these levels are preceded by one related to notions of organisational culture, environment or climate. The latter is clearly of significance at a time when the external environment of economics and politics is a major influence on changing organisational cultures.

In both public and private sectors the changing nature of client/customer expectations is a major driver. We can conclude, therefore, that organisational culture and externalities are mutually dependent.

We can also assume that, subject to the action of externalities, an organisational culture derives from corporate management and that its characteristics will be heavily dependent on its core business. In many cases FM has been responsible for initiating culture change in the operation of key aspects of the organisation's business, for example within the UK public sector generally and perhaps the health service in particular.

An understanding of organisational culture is essential if FM is to be effectively implemented and is part of a wider issue that ensures, as in a classic project management exercise, that the project is fully scoped.

As well as ensuring that all stakeholders have a proper grasp of organisational culture it is essential in a scoping exercise to ask such fundamental questions as:

❑ How can we work with the existing culture?
❑ Are there characteristics that can be turned to advantage in the strategy adopted?
❑ Does the existing culture impede development?
❑ If it does, do the solutions lie in initiating a culture change, or should the strategy be adapted to meet this challenge?
❑ Are the existing cultural characteristics consistent with external imperatives?
❑ Can the existing organisation effectively accommodate the changes that the FM operation will bring? Should it be done at all?

Any project of this type becomes a real exercise in change management, in which the principles of project management should be followed. Without robust project scoping at the outset, the success of the following levels may be seriously compromised.

At the strategic level, then, the task is to begin to build on a proper scoping exercise and engage in a carefully derived planning process. In strategic management terms it is normal to think in terms of planning within a three- to five-year time frame within which more detailed short-term plans are developed. This is consistent with the planned maintenance framework described later.

In the broader FM context, one of the other major outputs from the strategic planning exercise should be a clearly articulated mission statement. The production of this may well beg questions at a corporate level, and consistency with corporate mission statements is clearly necessary. Strategy development is something of an iterative exercise, and firming up of key objectives and movement to the next level of detail should not occur until the required number of iterations has taken place.

A mission statement should include a number of key statements defining the strategic objectives and these then provide a basis for detailed planning of actions that are more specifically detailed within shorter-term time frames.

An important output at this stage, which is a result of both corporate and strategic processes, is the formulation of an FM policy. Again it must be stressed that without proper scoping, strategy formulation and iteration a robust policy statement will not emerge. In developing FM policy at this stage a start can be made to:

❑ define roles and responsibilities
❑ produce a management organisation structure
❑ begin to define operating structures
❑ design communication systems
❑ 'sell' the proposition to all the stakeholders, many of whom have not yet been involved
❑ develop an embryonic performance measuring system.

The policy statement endorses senior management commitment to the project and this can be thought of as a Project Gateway[6] which has to be passed through before we move to the tactical phase.

At this level executive responsibility should be taken to put organisational structures in place and define management roles associated with:

❑ setting standards
❑ implications for managing change down to a departmental level
❑ identifying and managing resource implications
❑ budgeting
❑ communication systems.

Planning at the tactical level will need to be accompanied by detailed documentation to fully describe all the activities associated with the operation of the facilities management process or processes to be undertaken, which effectively becomes an operational manual. The documentation not only ensures a shared understanding amongst all parties; its preparation will also force management to ask all the essential questions necessary for a proper infrastructure to be put in place to underpin the operational level of FM.

At this stage important issues related to control systems, in terms of money, time and quality, and human resources management are identified and, significantly, an understanding of the manner in which the FM services are to be procured is developed. The literature is rather conflicting as to which level this resides in, and it is wise to assume that the levels referred to should be seen not as discrete but as a continuum from which the procurement strategy emerges as one of the key FM policies.

Early, rather simpler views of FM saw it as an internal organisational issue aimed at more effective structures within the organisation for managing its facilities. However, the evolution of facilities management has seen a gradual shift towards a position where FM services are contracted out or outsourced.

Although outsourcing and facilities management are often seen as inseparable, it is stressed that they should in fact be seen as distinct processes, with the former being one of a number of ways in which a facilities strategy can be implemented.

Public sector developments

The pace of change in the public sector is rapid, particularly in local government and, for example, in the NHS where notions of value and performance are driving an agenda that means building maintenance is being executed within a very different context than hitherto. Public sector organisations pose interesting questions as to what represents their core business. In some cases, such as that of local authorities, this question is complicated and answers may tend to be somewhat politically loaded. On the other hand, the NHS clearly defines its core business as medical and clinical patient care and the ancillary services associated with this, such as laundry, cleaning, building maintenance and catering, can be contracted out or outsourced as part of its facilities management strategy. Increasingly, in fact, as a cursory study of the websites of service providers reveals, outsourcing is becoming almost synonymous with facilities management.

Although the core business of a local authority is less easy to define, around the edges it can contract out or outsource a number of functions. Refuse collection, for example, is commonly subject to this process.

All public sector bodies do, however, have a set of common characteristics. There will be:

❑ multiple service objectives and multiplicity of expectations for each of these
❑ a number of external and internal influences on policy – economic, social and political
❑ strong influences from funding bodies
❑ a possible tension between local and national funding agencies
❑ a high proportion of resources from central government
❑ resources maybe received 'up front' but with constraints on how they are utilised
❑ recipients of the service who do not directly or in some cases even indirectly pay for the service.

Managing an estate in this context is therefore complex and the development of a strategy has to be carried out in the context of major pressures and strong competition for resources. What is certain, however, is that poor management of the estate will have a detrimental effect on the delivery of other services. Unfortunately it is not often recognised that a well-run estate contributes to improved quality.

Given the range of pressures, it is not surprising that FM has only recently been seen by much of the public sector as an all-embracing integrated approach.

Alexander[7] states that facilities management in the public sector needs to be:

managed as a public service [and] balanced with the democratic needs of the local community. Public services are managed for social results and must maintain a strong financial performance while achieving primary social goals such as equity, access, community accountability, environmental responsibility and equal opportunity.

He goes on to say:

Facilities Management in the Public Sector is not only about organisational effectiveness, but also about public service and responsibility, particularly in the face of change. The challenge is to bring about change, to anticipate public needs, to identify the contribution that facilities make to public service, and to manage for the effective use of public assets and resources.

Implicit in this is that the operation must deliver value for money and it is here that perceptions of value give us cause for concern. There are numerous attempts to define this in facilities management terms. For example, Featherstone and Baldry[8] offer:

The process by which the facilities function creates and nurtures an optimum environment and delivers effective support services to meet organisational objectives at best cost.

It becomes clear from this that for a public sector organisation to develop an effective FM strategy there has to be a comprehensive understanding of the climate/environment within which it operates, and a clear recognition that within this there has to be a customer/public focus. Although this can be said to be true also of a private sector organisation, one has to keep a clear awareness that the customer for the public sector organisation is the public at large, and a very different set of perceptions will operate. Within this environment successive governments have sought to increase the effectiveness of public sector service delivery, often with the focus very firmly on the cost rather than the quality side of the value equation. To raise awareness of value in terms of public services is therefore not easy.

Local government

Within local government the dominant theme at the time of writing is one of 'best value'. The Local Government Act 1999 placed a duty on local authorities to:

Make arrangements to secure continuous improvements in the way in which they exercise their functions, having regard to a combination of economy, efficiency and effectiveness.

Local authorities are required to set measurable standards for the services they manage and undertake performance reviews on these services over a five-year period. These reviews are required to demonstrate that continuous improvements are being made.

In 2000 'best value' replaced the previous regime, which relied on Compulsory Competitive Tendering (CCT) to improve performance and cut costs in the public sector. The stated aim of the best value initiative was to:

ensure that within five years all council services achieve performance levels that were only achieved by the top 25% of councils at the start of the five-year period.[9]

With the emphasis on continuous improvement and performance monitoring, the major responsibility for administering the regime lies with the Audit Commission. Opposition to the initial scheme focused around its heavy-handed approach and a resource-hungry, time-consuming inspection procedure. As a consequence of this, the system was modified so that the best-performing councils should only be subjected to a light-touch approach in terms of auditing. This led to the introduction of the Comprehensive Performance Assessment (CPA) via a White Paper in December 2001. This comprehensive inspection was aimed at ranking all councils into categories 1 to 5, poor to excellent. As an incentive, councils with the top rankings would be given increased autonomy in terms of their budgets.

Under this evolving system councils were now no longer required to produce detailed reviews of all their services but only those highlighted by the CPA.

Best value is, however, the underlying theme throughout all this evolutionary process and the principle is that councils should apply what are called the four Cs:

❑ Challenge
❑ Consult
❑ Compare
❑ Compete.

Under the Act councils produce a Best Value Performance Plan (BVPP) within which are embedded performance indicators (BVPI). There are two types of BVPI:

❑ Best Value corporate health indicators to provide a snapshot of overall performance
❑ Best Value service delivery indicators.

Up until 2002/3 the Audit Commission also specified performance indicators. This is no longer the case as the government continuously seeks to streamline the best value processes. Although best value replaces CCT, this forerunner was, and still is, influential in that it encourages local authorities to take a more holistic view of performance, and this includes the manner in which it procures its services. This has inevitably led to an FM approach and an increased tendency to contract out or outsource. It has also begun to seriously challenge what were hitherto rigid approaches to procuring those services and built assets, particularly schools, although it must be pointed out that there are also other drivers for the changes in procurement practices within local authorities.

The National Health Service

The creation of the internal market in the National Health Service in the 1990s was a major catalyst for change, which forced NHS trusts to examine very critically the way in which non-clinical services are delivered. The response of many trusts was initially to create single directorates to bring together estates and support services such as laundry, catering and cleaning.

Over the last ten years the NHS, FM and the notion of the internal market have evolved quite radically. The White Paper 'A New NHS'[10] shifted the emphasis away from competition towards partnership and collaboration. The philosophy embedded in the White Paper emphasises that quality of service should be based not only on clinical treatment, but on the whole healthcare experience. Improving the quality of support services is therefore seen as an integral part of continuous improvement and encourages a general move towards a strategic facilities management approach. Implicit in this development have been two key trends:

❑ outsourcing of services considered to be supporting core activity of clinical care – often through an integrated FM package

❑ for the development of new facilities, a Public Private Partnership approach through the Private Finance Initiative.

Both of these issues merit particular attention as they are of huge significance to the context within which maintenance is carried out, not only in the health service but also in the rest of the public sector and in the private sector.

Outsourcing

Outsourcing has been defined as a process whereby an organisation employs a separate company or supplier to perform a function that had previously been carried out in house. This is often accompanied by a transfer to the contractor or supplier of assets, including people and management responsibility.

The objective of outsourcing is for an organisation to divest itself of non-core, peripheral functions in order to reduce costs in performing and managing what are considered low-priority operations. Associated with this is the prospect of improving efficiency through the use of specialist providers. Within the public sector the search for increased efficiency driven by best practice has been a major stimulus to outsourcing.

Each organisation may develop its own view as to the services it wishes to outsource, depending on its definitions of core and non-core business. Deciding what is and is not core business has led to heated debate within organisations. However, four basic models have developed:

❑ A diverse range of services are contracted out piecemeal by a range of managers from various parts of the organisation. Although on the face of it this appears to lack a co-ordinated strategic approach, it may in fact represent a pragmatic view for some types of organisation.

❑ A centralised control of outsourced service contracts has, however, become a much more common approach and has many benefits in terms of achieving value for money from the exercise. Centralised monitoring of service quality fosters the development of expertise and better evaluation, with the potential to draw comparisons and even engage in internal benchmarking exercises and co-ordinated and shared data.

❑ The grouping together of a number of contracts is referred to as 'bundling'. This approach begins to move the organisation seriously towards a much more strategic view in terms of decision-making. There are various levels of bundling. At one extreme the collection of services having similar characteristics may occur to form a number of 'bundles'. For example, the grouping of cleaning and security seems sensible. At the other extreme, a single strategic bundle may be formed, and we move into the realm of total outsourcing, or total FM.

❑ Total FM places all service contracts under the direct control of an FM company. This creates an environment that has the potential to deliver very efficiently but clearly has attendant risks. To manage such a service requires sophisticated control and monitoring systems and requires a service provider with highly developed management skills. There is obviously a large amount of trust implicit in this approach, and partnering arrangements, particularly in the public sector, are becoming commonplace. The provider is unlikely to be able to offer the complete package of services in house and some elements may be subcontracted. This has the potential to create complications, particularly in respect of accountability.

Facilities management outsourcing experienced massive growth during the 1990s and it is estimated that in 2003 the market was worth approximately £12 billion.[11] This growth has been underpinned not only by a buoyant corporate sector, but also the emergence of Public Private Partnerships (PPP) arising from the government's drive to improve the quality of the public sector building stock, notably schools and health buildings where the Private Finance Initiative has been the main procurement tool.

The trend in outsourcing has been towards 'bundled' service delivery and ultimately total facilities management rather than piecemeal single-service packaging.

It is clear from all this that outsourced service contracts are complex. A policy of renewing contracts at regular intervals is clearly tempting as it provides the option to rid oneself of a bad provider and keep the contractor on his toes, with the potential for cost-saving through rigorous competition. However, careful account should be taken of the nature of the work. With work of an ongoing nature, the benefits of building up lasting working relationships with a provider are not to be taken lightly. There is thus a balancing act to be achieved and the need for an extremely careful approach to the selection process, robust specification of service needs and a fully shared understanding of what is required. This is an extremely complex task.

When seeking a contractor, it is imperative that the client has a clear specification and that this is accompanied by a clear explanation of what is required of the contractor. This is expensive and time-consuming but should culminate in a clear, unambiguous set of actions, outcomes and quantifiable performance requirements. It is essential, therefore, that the rules of engagement are clearly defined and, as what is being outsourced is clearly a service, careful consideration of these needs to be framed into a service level agreement. By implication these agreements lead to the establishment of relationships that are longer term than those existing in a more traditional client–contractor engagement. In the context of the public sector this inevitably draws us into consideration of partnering and the drive from central government as reflected in a number of publications that emanated from the Office of the Deputy Prime Minister* targeted at improving service delivery.

---

* At the time of compiling the text for this book, the Office of the Deputy Prime Minister was responsible for driving forward many of the agendas referred to in this chapter and later ones. Between that time and going to publication, the Office of the Deputy Prime Minister was effectively disbanded and its responsibilities gradually re-distributed. At the time of writing this note the process is still ongoing. This presented us with something of a dilemma, but after consideration we have left the original reference sources intact. We apologise if in the future this provides the reader with difficulty in tracking down some of these sources. An attempt to do anything other than this would have been 'crystal-ball-gazing' to some extent, and would have been equally confusing.

Public Private Partnerships (PPP)

Traditional contracts tend to be very specific, with performance measured against tightly articulated requirements with defined penalties for non-compliance. There is a tendency for them to be self-perpetuating in the sense that they are framed in a manner that assumes things will be done in the same way as they were in the past. The contractual relationship does not therefore encourage flexibility and innovation and may degenerate into an adversarial relationship between the parties. On the other hand, a partnering arrangement seeks to:

> take a long-term perspective on needs, costs and solutions; involve shared risks, costs and rewards; have agreed problem resolution methods; have joint governance arrangements and [be] capable of change and development.[12]

Government publications refer to strategic partnering because it believes that real improvements can be created if the public sector in general, and local government in particular, takes a corporate view of long-term objectives rather than a piecemeal approach. Therefore in order to assist local authorities to develop effective partnerships to achieve its commitment to 'drive up the standard of public services',[13] the Strategic Partnering Taskforce was formed in September 2001. This group embarked on a programme of projects with 24 so-called Pathfinder Strategic Service Delivery Partnerships in order to identify and draw out good practice. The results of this work have been disseminated in a series of 'Rethinking Service Delivery' publications. These publications also draw on the work and guidance provided by a number of other initiatives and provide a basis for addressing the recommendations of the Audit Commission's report *Competitive Procurement*.[14] Thus the concept of Strategic Service Delivery Partnerships has emerged.

These initiatives all fall under the broad umbrella of PPP. They significantly influence the manner in which the maintenance of buildings is executed in the public sector and will continue to do so for the foreseeable future.

PPP is a generic term for a range of scenarios in which relationships are forged between public and private sector bodies with the aim not only of injecting private sector finance into the improvement of public services, but also of bringing together expertise from both sectors. The expectancy is that these partnerships will be increasingly effective in terms of both procurement of public facilities and ongoing service delivery. In some scenarios procurement, construction and operation are merged into one agreement. The working relationships set up can be loose, formal, short-term or long-term, or strategic. In its most advanced form a strategic partnership to design, build, finance and operate (DBFO) effectively creates a service contract and may be brought about by the formation of joint venture companies.

Amongst the variants is the Private Finance Initiative (PFI), the key characteristics of which are:

❏ provision of capital assets funded by the private sector partner
❏ a long-term service contract for a private sector partner

❑   a single unified payment from the public sector body
❑   integration of design, construction, finance and operation
❑   complex arrangements for the allocation of risk
❑   an arrangement for the specifying of performance standards for the service delivery
❑   performance-related remuneration for the private sector partner.

Since its introduction by the Conservative government in 1992, use of PFI has accelerated markedly. For example, the NHS plan, at the time of writing, states that 100 new hospitals will be delivered under PFI by 2010.

By 2005 more than 400 PFI projects, each having a capital value over £15 million, had been signed. Given the nature of the service delivery package, PFI is having a profound effect on the way in which maintenance services are delivered, although the full significance of this is not yet fully understood other than anecdotally. That the approach is shrouded in controversy is underlined by an article in *The Guardian* which said:

> Growing concern has recently been expressed amongst experts about the cost of PFI. Public Sector Accountants claim that hospitals and schools would be cheaper to build using traditional funding methods. The National Audit Office described the value for money test used to justify PFI projects as 'pseudo-scientific mumbo jumbo'.[15]

However, according to the National Audit Office,[16] 78% of projects have been delivered on time and at agreed cost. Very tellingly, with respect to the implications for building maintenance, the *Guardian* article quoted above concluded with the comment: 'However, it will take at least another 20 years, when the first PFI contracts have been completed, before the real cost [or value] of PFI can be judged.'

Despite these concerns, the present government remains committed to the PPP concept as it recognises that the public sector working in isolation cannot finance the investment required to improve the nation's infrastructure. The use of PPP service delivery is therefore likely to remain on the agenda for the foreseeable future and it is now clearly acknowledged that new skills and attitudes are required to establish a proper framework for its implementation.

The Strategic Partnering Taskforce is central to this agenda, and in particular its guidance on Strategic Service Delivery Partnerships. It defines a Strategic Service Delivery Partnership (SSP) as:

> A long-term partnership between organisations that work collaboratively to achieve the authority's aims for delivering services.

There is an emphasis on the partners developing a high level of shared objectives, and to achieve best value for money coupled with a culture of continuous improvement. By implication this should generate a constructive, genuinely collaborative relationship.

It should be emphasised here that philosophically the objective is to generate partnerships, and in terms of maintenance management it is the PPP environment that is likely to have the most significance. The Task Force identifies the following key aspirational characteristics that distinguish an SSP from a conventional contract:

❑ Detailed contracts are used but there is a greater focus on relationships and achieving compatibility of management cultures.
❑ Management and administrative protocols are developed in a collaborative environment.
❑ A methodology for evaluating/measuring outcomes and achievements is agreed.
❑ There are equitable payment mechanisms.
❑ Monitoring mechanisms are developed by the partnership that are transparently open to external audit.
❑ A culture of problem solution is emphasised in place of the 'blame culture' that often characterises conventional contracts.
❑ Re-evaluation and monitoring systems are developed, aimed at continuous improvement (improvement planning and value engineering).
❑ There is a shift away from a lowest cost philosophy towards one of value and quality of service delivery.
❑ The partnership have an agreement to a flexible approach to achieve continuous improvement.

A number of issues are raised by these characteristics, and research on the operation of PPPs suggests there is still some way to go before they are all resolved. Of particular interest is the issue of management culture and the need to achieve alignment of partners' objectives.[17] Whilst the contract can define the limits of each partner's responsibility, it does not specify deliverables without which the main thrust of the aspirations listed above cannot be achieved. A wide range of management issues are acknowledged, and will only be resolved with increased experience and enlightened and robust staff development. To ensure shared understanding and to fully elaborate on deliverables, supporting infrastructure documents need to be in place.

A key document is a service level agreement which, as the name implies, defines expected levels of performance. Defining performance levels is a related generic issue in its own right in all partnerships, contracted-out or outsourced arrangements, and these receive more specific attention later in this text when maintenance execution issues are explored.

References

(1) Gribben, R. (2003) *Daily Telegraph* news article.
(2) Booty, F. (2003) *The Facilities Management Market.* RICS, London.
(3) www.msi-marketingresearch.co.uk
(4) www.bifm.org.uk
(5) www.facilities-centre.com
(6) www.ogc.gov.uk
(7) Alexander, K. (ed.) (1995) *Facilities Management.* Blenheim Business Publications, London.

(8)   Featherstone, P. & Baldry, D. (2000) *Public Sector Facilities Management Strategy: Construction and Building Research Conference.* Greenwich University, London.

(9)   Weaver, M. & Parker, S. (2002) Best Value and Inspection: The issue explained, *The Guardian* 29 October.

(10) www.nhs.gov.uk

(11) www.amaresearch.co.uk

(12) Office of the Deputy Prime Minister (2005) *What is Strategic Partnering.* ODPM, London.

(13) Hope, P. (2003) *Rethinking Service Delivery.* ODPM, London.

(14) Audit Commission (2001) *Competitive Procurement.* Audit Commission, London.

(15) Weaver, M.(2003) PFI The Essue Explained, *The Guardian*, 15 January.

(16) National Audit Office (2003) *Construction Performance PFI.* National Audit Office.

(17) Wagland, G. (2000) 'Public Private Partnerships'. Unpublished MSc dissertation, University of Bath.

Chapter 2
The Maintenance Dimension

Building maintenance has consistently been treated as the 'poor relation' of the construction industry, attracting only a tacit recognition of its importance, both within the industry and amongst building owners. This manifests itself in a general lack of understanding of both its scope and its significance by all parties to the building procurement, construction, and management processes. In consequence, the backlog of repair and maintenance work required to bring the country's building stock to a minimum acceptable level continues to grow at an unacceptable rate.

Latterly, for a number of reasons, many of which are discussed in the preceding chapter, the dimensions of the problem have forced it higher up the agenda and promoted what appears to be greater professional interest. This increasing level of concern over the condition of the nation's building stock has served to expose more clearly the extent of the problem. Whilst effective maintenance policies are not by any means the norm, the efficient utilisation of scarce resources is beginning to be approached in a more informed way, and the fundamental relationship of the condition of a building's fabric to its total performance examined more critically.

Maintenance defined

BS 3811: 1984 defines maintenance as:

> A combination of any actions carried out to retain an item in, or restore it to an acceptable condition.[1]

From this definition two key components can be identified:

- ❑ not only actions that relate to the physical execution of maintenance work, but also those concerned with its initiation, financing and organisation
- ❑ the notion of an acceptable condition, which implies an understanding of the requirements for the effective usage of the building and its parts, which in turn compels broader consideration of building performance.

The latter presents some problems when attempting to determine the standard that represents an acceptable condition, as opinions will vary from person to person and over time according to the type of building under consideration, its usage and changing circumstances. Within the private sector, for example, there are likely to be quite different opinions as to what an acceptable condition is between parties on opposing sides of a tenancy agreement. These perceptions may range from a 'cosmetic' view to an in-depth evaluation of building performance needs, and will certainly be fluid enough to drift considerably with changing market conditions.

Within the public sector, it is apparent that social and political forces will have some bearing, and it may be extremely difficult to justify a given stance in the same terms that might be exercised in the private sector. The nature of the responsibilities of the public sector highlights the need to deal not only with the satisfaction of individuals, but that of groups of people and society in general.

Acceptable condition can also be looked at in quite another way by trying to view the physical needs of the building in an objective manner, although even then problems of perception will be just as prevalent. It is only necessary, for example, to contrast the different attitudes taken to fabric maintenance and the maintenance of engineering services to see this demonstrated. Acceptable standards will also change with time, and given the long life of buildings this is of major importance.

The Committee on Building Maintenance (1972) recommended the adoption of the following definition of maintenance:

> work undertaken in order to keep, restore or improve every facility, i.e. every part of the building, its services and surrounds to a currently acceptable standard and to sustain the utility and value of the facility.

This definition is, if anything, rather broader than that in BS 3811 as it introduces the notion of value, which is linked with life expectancy and requires consideration of the complex mechanisms that either erode or enhance the value of a built asset over time.

The usefulness of both definitions will depend on whether an economic/financial appraisal is being undertaken, or if the real concern is with operational and building condition issues.

The question also arises as to whether or not 'improvement' should be included under the heading of repair and maintenance. In any maintenance operation there will almost always be an element of improvement, if only through the replacement of an obsolete component. Thus to a large extent maintenance and improvement are inseparable, but in principle are clearly distinguishable from conversion, rehabilitation and refurbishment, which have the clear objectives of adapting or increasing the utility of a building, rather than maintaining it at the current level. This distinction between maintenance and improvement is of more than passing academic consequence as it may have a significant influence on funding arrangements. Housing associations, for example, have until recently been able to attract grant aid for major improvements, whilst maintenance was funded from revenue.

The most sensible approach to take at the outset is to see maintenance work as that which enables the building to continue to efficiently perform the functions for which it was designed. This may include some upgrading to raise the original standards, where appropriate, to contemporary norms and the rectification of design faults. Thus building maintenance needs to be seen as a part of a larger property management function and viewed in the context discussed in Chapter 1.

Both definitions of maintenance imply an interest in actions, which may be either physical or organisational. It is with the latter that this text will be primarily concerned and so it is necessary to consider maintenance within the micro-business environment of the company or organisation. Because building maintenance operates within a complex macro-climate, the level of activity will be determined by a host of interrelated factors. These can broadly be categorised into:

❑ supply side factors, concerned with the ability of the industry to execute defined maintenance requirements
❑ demand side factors, associated with the definition of maintenance tasks and the ability and willingness of building owners to execute them.

Demand for construction work

The determination of demand for construction activity is a very complex affair and heavily dependent on the policies of government, either directly through its intervention in the public sector, or through its influence on the general level of economic activity in the private sector. To the economist, demand is regarded as the requirement for goods and services that the customer is able and willing to pay for. It is thus important to distinguish between demand and need.

In terms of maintenance, need may be measured by reference to the standard of building condition that is regarded as being acceptable, and the extent to which the actual condition falls below this standard. At any one time there may be a large quantity of latent demand awaiting the right conditions for its realisation. These conditions will more often than not be financially related.

Hillebrandt identifies four major requirements for the creation of demand:[2]

❑ the existence of a user, or potential user
❑ someone willing to take on ownership responsibilities
❑ available finance
❑ the existence of an initiator, such as a professional building manager.

In addition to these factors being present, it is also necessary for contextual conditions to be favourable; thus external agencies, the most important of which is government, will be extremely influential in determining when they arise.

The supply of construction services

From time to time high levels of demand expose an inability of the industry to deliver the service required by its clients. This normally manifests itself in labour shortages and extended delivery dates. The reasons why this occurs are complex, but appear to stem from the cyclical nature of the construction economy, a lack of responsiveness to the cycles, and difficulty in attracting talented people into the industry. Such times are characterised by inflated tender figures and a tendency to seek less labour-intensive ways of building, sometimes with unhappy results: for example, the system-building housing schemes of the 1960s.

Demand for construction work is always extremely varied and the characteristic structure of the industry, with its diversity in size and type of company, has evolved in order to supply this diverse demand. When demand is low the industry attempts to adjust and contractors will compete outside their normal markets. This may lead to their involvement in work for which they are not necessarily suitable. Periods of recession lead to a slimming down of the workforce and there is a tendency in construction for good-quality labour to be lost permanently, which in turn tends to exacerbate inflationary problems in an upturn.[3]

The structure of the construction industry

The construction industry has many characteristics which, taken on their own, are not unique in industry as a whole. The peculiarity of construction, compared with other industries, stems from the existence of a unique combination of these factors.

The industry is one of paradox in that, on average, its annual output contributes approaching half of the fixed capital formation of the country, and yet it is characterised by the presence of a very large number of small companies. When the size of the construction industry is measured in terms of its contribution to the gross domestic product (GDP), it can be seen that, even in a lean year, this is around 10%.

Its product is large, heavy and immovable, ranging from small-scale housing and extensions, through large sophisticated office and hospital buildings, to vast civil engineering projects. Most of its output is custom-built and geographically widely distributed. In terms of production, its shop floor is a constantly changing one, and is characterised by the movement of men, machines and materials to the emerging product, rather than by production-line processes. It therefore suffers more difficulty in mechanisation and rationalisation than most other industries.

The output of the industry is extremely variable, and the idea of construction being used as the regulator of the economy is well known. However, as the influence of the public sector has diminished in terms of direct output, this is no longer the case. There is also a suggestion that the cyclic nature of output over the last ten years is less pronounced. Figure 2.1 shows output over the period 1994 to 2002 indicating steady growth but with fluctuations in the rate of growth.

Another important measure is the number of people employed. This fluctuates either side of 1.5 million, accounting for 6–7% of the workforce. The number of people employed evidently fluctuates with the industry's output. However, there is evidence of a decline in real terms. Of particular concern for maintenance operations is an adequate supply of well-trained labour.

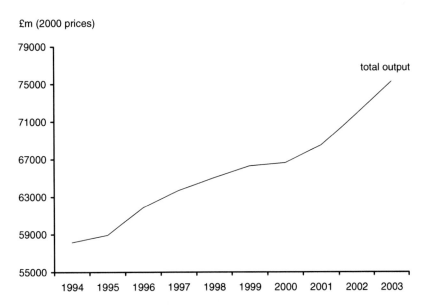

£m (2000 prices)

Figure 2.1 Construction industry output. *Source*: DTI

No other industry is more fragmented or diverse in its structure, and it is still easier to form a construction company than almost any other. Figures are notoriously unreliable, but DTI Construction Industry Statistics 2004 suggest there are in the order of 170 000 construction companies, of which in the region of 80% employ fewer than 14 persons whilst 0.16% employ more than 300 personnel. In 2003 there were in excess of 70 000 single-employee companies and only 64 employing more than 1200[4] workers.

Building maintenance in the construction industry

Supply and demand

To create demand for maintenance activity there is a range of factors, which will be discussed later in this chapter. For the time being, however, it may be assumed that the general environment favourable for demand creation is as relevant to maintenance work as it is to construction activity in general, although the mechanisms at the micro-economic level may be different.

The influence of the following factors in particular should be considered:

❑ The general economic climate and its trends in respect of:
  ○ interest rates
  ○ inflation
  ○ industrial output
  ○ exchange rates.
❑ The cost of construction in relation to the cost of other goods or services.

❏ The operation of the rules and regulations governing land use and property development.

Within the construction industry, the various sectors compete with each other for resources. This is true of maintenance activity, which, because of the way in which it is viewed by the various parties, may be at a relative disadvantage in this respect. Additionally, maintenance has to compete for resources in another market, that of the general business environment. Within both markets the supply of financial resources is clearly important, but within the construction industry itself there should also be some concern with regard to the supply of physical resources, such as an adequate and properly trained workforce. This is a complex issue, and is dependent on the levels of available employment opportunity generally, and hence the level of demand for construction activity in particular.

Maintenance output

Data on maintenance output can be obtained from the construction industry statistics, which separate out repair and maintenance work from other construction output. However, these statistics need to be treated with caution, as the work of maintenance departments is not always clearly identifiable because of the complexities of definition.

Figures from DTI statistics for 2004 indicate that total expenditure on maintenance in the UK increased by 12.2% from 2002 to 2003 and by almost 90% in the ten years from 1993.

Figures 2.2, 2.3 and 2.4 show repair and maintenance output in comparison with the total output of the industry. Between the 1950s and 1982 repair and maintenance,

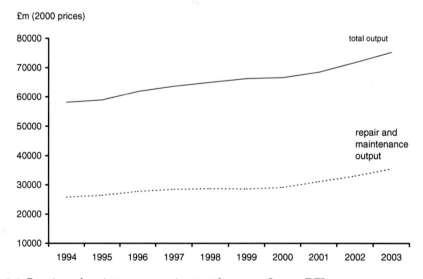

Figure 2.2 Repair and maintenance against total output. *Source*: DTI

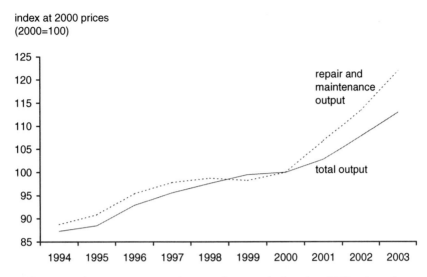

index at 2000 prices
(2000=100)

Figure 2.3 Repair and maintenance against total output indexed at 2000 prices. *Source*: DTI

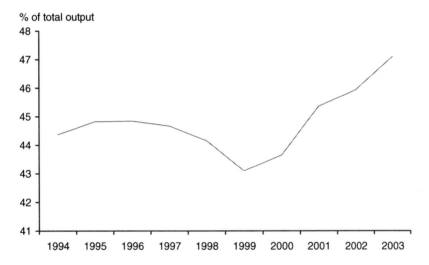

% of total output

Figure 2.4 Repair and maintenance as a percentage of total output. *Source*: DTI

including housing improvement, increased its share of construction output from 25%
to around 47%. Since then, there was a dip at the end of the 1990s but steady growth
from 1999 to around 47% in 2002 (figure 2.4).

These percentage figures on their own do not of course give the full story as, even
at a constant level of maintenance activity, they will rise if overall output declines. For
example, the fact that historically much of this increase occurred in a period of recession for the industry as a whole has prompted claims that the industry's resources are

switched to this sector when new work is hard to come by. This claim is difficult to fully substantiate, not only for the reasons given above, but also because both supply and demand factors are influential.

It should be noted that, when the period considered above began, there was still a substantial shortage of buildings as a result of World War II. By the mid-1970s this demand was slowing down, whilst at the same time the availability of land for new building was under pressure. More recently, there has been increased dissatisfaction amongst the general public with much new building, which has focused attention on the better use of the existing building stock. Thus, although a substantial proportion of the stock has become unfit for the use for which it was designed and built, this has created demand for adaptation and refurbishment. These pressures on the realisable demand for new work tend to increase the latent demand for maintenance work in order to extend the life of the asset, even though general economic conditions may not be favourable. Any changes in the proportion of maintenance work over this time should also be judged against a low base level, representing a very minimal level of maintenance activity at the beginning of the period.

In the private sector, because the expenditure on existing buildings is likely to be internally financed, the user, owner, initiator and financier may be the same. The property industry may differ, but at this stage it is assumed for the sake of simplicity that free market pressures on rents will bring user and owner into harmony. The decision-making process is therefore likely to be different from that in the public sector. Much maintenance work in the private non-residential sector is thought to be determined by profit levels, and from time to time these may be under pressure.[5,6]

In the public sector, the level of maintenance activity among the various parts is worth examining, as each of these is the subject of strong political pressures from time to time, which can distort the distribution of maintenance activity. Housing is an obvious example, and the need for improvement in this sector has been well documented in the English House Condition Surveys (EHCS), although there are doubts as to whether these produce a genuine picture of reality, as discussed later.[7] The publication of the EHCS report back in 1981, for example, led to a massive political outcry and increased activity in maintenance work on housing, which may have had the effect of increasing maintenance output, or led to a switch of finance from other areas such as schools maintenance. The picture is, however, currently more complex with public sector changes of ownership.

Figure 2.5 shows a breakdown of repair and maintenance work for the period 1994 to 2002 and figure 2.6 shows public housing maintenance as a percentage of total housing maintenance declining over the period. Figures for housing maintenance as a proportion of total repair and maintenance are given in figure 2.7, and this also shows a decline.

When examining maintenance statistics, it should be noted that the figures given by DTI include a contribution for maintenance work carried out by directly employed labour in private sector organisations, and for DIY housing maintenance.

As a percentage of GDP, maintenance output is rather erratic as indicated in figure 2.8 although a sharp rise since 2000 is evident. A comparison of maintenance

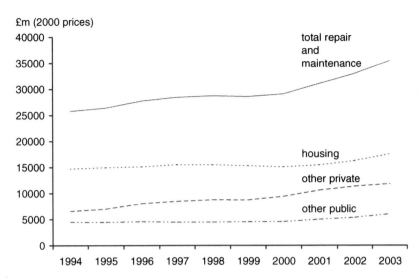

£m (2000 prices)

Figure 2.5 Breakdown of repair and maintenance output. *Source*: DTI

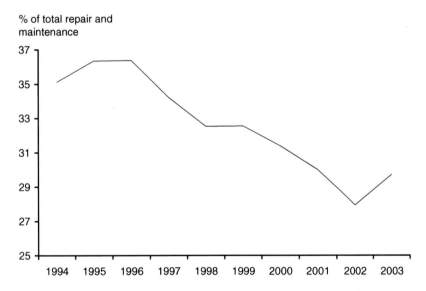

% of total repair and maintenance

Figure 2.6 Public housing as a percentage of total housing maintenance. *Source*: DTI

expenditure in relation to the replacement value of the building stock, using data from BMI, in figure 2.9 also shows no clear trend. However, the figures do show that housing maintenance expenditure, in relation to the value of the stock, has been consistently higher than for either public or private non-housing, although the figures do include some alteration and improvement work. If the estimated value of DIY labour is included,

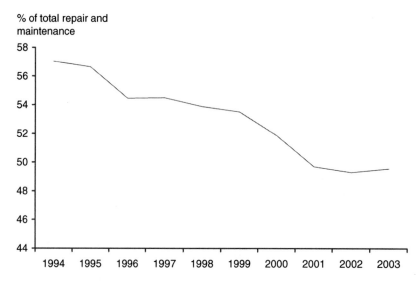

Figure 2.7 Housing as a percentage of repair and maintenance. *Source*: DTI

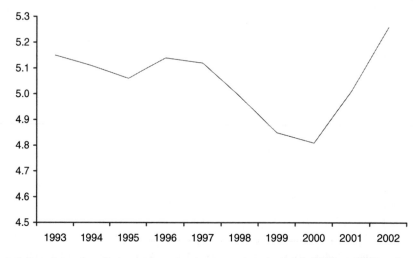

Figure 2.8 Repair and maintenance output as a percentage of GDP at 1995 prices. *Source*: BMI

the percentage rises by around half a percentage point. Whilst it must be borne in mind that stock valuations are rather imprecise, the trend appears to be clear. The raw statistics also say little about the condition of the building stock, for example its age profile, and there is a strong case for arguing that maintenance expenditure, expressed as a percentage of stock value, should be increasing when this is taken into account.[8]

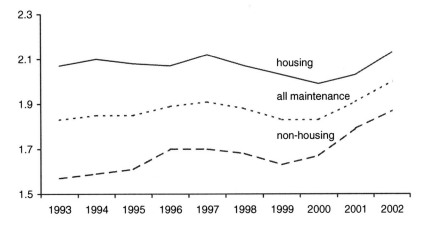

Figure 2.9 Repair and maintenance output as a percentage of capital stock at 1995 prices.
Source: BMI

Government policy and maintenance

Since 1979 the role of government, in terms of economic management, has moved away from Keynesian principles, with their emphasis on demand management, and returned to the classical theories espoused by Adam Smith and others. Government thus adopts a far less interventionist approach and relies on supply and demand controlling the market, with government using indirect means of influence such as money supply and fiscal measures. Direct control of interest rates has now also been removed from government and rests with the Bank of England. It is useful, therefore, to consider the influence of some of these factors on the level of maintenance activity.

Taxation

This is of some significance, since maintenance expenditure by businesses is classed as revenue expenditure and is allowable against income and corporation tax, whereas capital expenditure is not chargeable on commercial buildings. It should also be noted that repairs attract VAT. It has been argued that this does not encourage an increase in capital expenditure to reduce maintenance, as the real cost of maintenance is reduced by the tax allowance. Conversely, it can also be argued that if a firm saves on capital expenditure, it may as a result increase its tax liability, through interest either earned or foregone by reduced borrowing. Movements in the level of taxation, however, will influence expenditure on buildings already in use, and there is a view that at higher levels of taxation increasing revenue expenditure to reduce tax liability is attractive.

However, the savings that can be made in this respect are heavily dependent on profit levels against which outgoings can be set.

Interest rates and inflation

The control of inflation has been the cornerstone of government economic policy for some time and this in turn is a major determining factor when interest base rates are set. Inflation is measured by the retail price index (RPI) and a monetarist policy on this implies the use of interest rates as the controlling mechanism. The theory is that in times of high inflation, high interest rates are used to bring it down. If this succeeds then lower interest rates may be used to stimulate the economy.

Following this economic theory it might be assumed, therefore, that low interest rates will encourage maintenance expenditure. However, there is little concrete evidence to suggest that this is the case. One reason for this is that maintenance expenditure only yields an indirect return, which is not readily represented by cash gain except, perhaps, if the property is to be sold. In the main the return is a long-term one.

Capital projects are far more responsive to changes in interest rates because of the effects on financing costs. High interest rates encourage the reduction of capital expenditure, perhaps at the expense of increased expenditure in the longer run. It is difficult to track this cause and effect cycle and thus this hypothesis has not been completely proven.

Economic confidence

Future expectations within the economy may affect maintenance expenditure, and industry looks to government, and other predictive indicators, for a lead in this respect. The building owner will thus set future investment levels and budgets in this context, and certainly within the private sector this will affect maintenance budgets and hence real expenditure.

Maintenance needs

Demand factors generally

Demand for maintenance output is subject, with some minor differences, to the influences identified above, and there is undoubtedly a significant latent demand awaiting the right conditions for its realisation. In the first instance, it can be generally recognised that there exists a need for maintenance work to put and keep the building stock in an acceptable condition.The precise extent of this requirement is only measurable in general terms. There are two reasons for this:

❏   a lack of universal agreement as to what represents an acceptable condition
❏   the absence of a fully comprehensive picture of the present condition of all our buildings, although reports focusing on particular building types are published from time to time, especially for the public sector.

At any one time some of this need for maintenance work is being satisfied. How much work is carried out reflects the willingness of clients to buy this service and their ability to do so.

It is important to distinguish between the ability to pay and the willingness to do so. To create maintenance activity both must exist. There may be a perceived lack of maintenance work, because of an inability to pay resulting from unfavourable economic or business conditions, or because a building owner is unwilling to buy even when finance is available.

A body of opinion exists that considers there to be a general lack of concern to repair buildings and maintain them in an acceptable condition. This under-resourcing is presumed to be the result of a massive level of ignorance and/or apathy. These rather general observations are supported by a number of reports, notably by the National Audit Office, that focus on the maintenance of the building stock in various parts of the public sector.

The influences and developments outlined in Chapter 1 are influential, particularly in the way in which the public sector building stock is managed.

Housing

DTI statistics indicate that housing maintenance represents close to 50% of repair and maintenance expenditure. This figure excludes a substantial input from the owner-occupier sector, especially that executed on a DIY basis.

The housing stock can be divided into the following categories, with the approximate percentage accounted for by each indicated:

- ❏ local authority (13.2%)
- ❏ registered social landlords (6.6%)
- ❏ private rented (10.4%)
- ❏ owner-occupied (69.8%).

There is other housing, notably that associated with the armed forces, for which detailed data is not readily available. An interesting point, worthy of note here, is that local authority housing has fallen from a figure of 22% in the early 1990s.

Whilst the National House Condition Surveys, National Audit Office reports and government websites provide valuable data on the condition of some of the housing stock, there is still some uncertainty about data on owner-occupied properties and those in the private rented sector owing to collection issues. This makes informed comment on housing in those areas rather difficult, and only generalisations can be attempted.

Between 10 and 15 million people live in what may be termed social housing in England and Wales, and the quality of their lives is directly affected by the condition of their homes and the environment of the estates in which they are located. Housing represents not only an important social and political issue, but also an economic one, when one considers that the replacement cost of the stock is in excess of £100 billion. An interesting perspective is obtained if one examines the history and development of the National House Condition surveys back to 1981. It paints a somewhat confusing picture but does underline the scale of the issue of housing maintenance.

The National House Condition Surveys back in 1981[9] and 1986[10] presented a very bleak picture of the condition of the stock and had the major effect of placing housing maintenance high on the political agenda. There is a suggestion, statistically, that these did have some effect in raising the profile of housing maintenance within the construction industry.

In 1986 the Audit Commission published two reports, *Improving Council House Maintenance*[11] and *Managing the Crisis in Council Housing*,[12] in which they described the scale of the problems facing many housing authorities, particularly those in urban areas, and notably Inner London. The crucial element of the problem was identified as the backlog of maintenance, repair and improvement work necessary to bring dwellings up to an acceptable standard. This backlog was estimated, in reports by the Department of the Environment (DoE), to be £20 billion in England alone, despite an average annual expenditure of about £425 per dwelling.

There were notable variations on the condition profile identified in the report owing to a variety of factors, one of which was relative age profiles, and another the extent to which backlogs were a function of previous underspending.

A large part of the problem was associated with the system-built dwellings of the 1960s and early 1970s, the majority of which were concentrated in urban areas.

Whilst the general thrust of these Audit Commission reports was directed towards the need to improve management practice, so as to make more efficient use of resources, they acknowledged that additional sources of funding would be necessary. It was concluded, though, that improved management could generate an estimated 30% in improved maintenance value.

There was a strong criticism at the time for changing the methodology between the 1981 and 1986 reports, particularly with respect to the definition of disrepair. It was also argued that the findings were not based on a large enough sample.

To aid comparison, the DoE carried out a smaller, separate 1986 survey of the dwellings surveyed in 1981, using the same methodology as the 1981 survey. This indicated that the number of houses lacking basic amenities reduced from 5% to 2.9%, but showed there was only a small decrease, from 6.3% to 5.6%, in the number of unfit properties, and from 6.5% to 5.9% for those in serious disrepair.

This survey also showed that in the 1981–1986 period only 42000 houses classed as unfit were cleared or closed, out of a total number of 900000 in 1986.

It concluded that poor housing continued to be occupied by low-income families; the proportion of poor housing was higher in rural areas (22%) than urban areas (14%) and, regionally, the North, Midlands and South-West fared worst.

At the time local authorities owned over half of all post-war dwellings in poor condition, and the DoE survey concluded that this was a result of poor initial quality and neglect in terms of planned maintenance. It also revealed that many authorities did not have accurate records of the condition of their properties, which is essential if proper programmed maintenance is to be carried out.

Private landlords make a relatively small contribution to the repair and maintenance of their dwellings, and the average cost of outstanding repair work is twice that of any other tenure. Another particularly hard-hit sector comprises houses in multiple occupation, and here the problem is compounded by inadequate or inconsistent data on the number of families in such accommodation.

The house condition surveys take account of all housing, but there is less substantive data available for owner-occupied stock. The DoE survey pointed out that the condition of owner-occupied dwellings gave cause for concern, and that owner-occupiers do not invest sufficiently in maintaining the condition of their property. There is some evidence suggesting that, in this sector, there is a substantial level of neglect, probably largely due to ignorance. In recent years there has been a steady increase in the proportion of owner-occupiers, particularly amongst lower-income families, and concern has been expressed at this, as many of the dwellings in this group may have been in questionable condition at the time of purchase.

In 2000 a government Green Paper[13] proposed the following:

❑ to bring all social housing up to a decent standard by 2010, supported by additional measures
❑ to ensure that local authorities adopt a more business-like approach to housing management and investment
❑ to support the transfer of up to 200000 homes each year from local authorities to registered social landlords (RSLs) where tenants vote for it
❑ to encourage local authorities to set up new arm's length management arrangements
❑ to promote and fund new PFI schemes
❑ to ensure the consistent and rigorous application of Best Value and Tenant Participation Compacts.

The proposed strategy targeted extra funding 'to tackle the £19 billion backlog of renovation and improvements'.

Four choices were made available to authorities and their tenants:

❑ Authorities retain ownership and management – extra funding through a Major Repairs Allowance to supplement other funding sources and their own revenue accounts.
❑ Set up arm's length arrangements – extra money on offer for this (outsourcing?) arrangement where excellence under the best value regime can be demonstrated plus other criteria.
❑ Provide extra funding for PFI schemes.
❑ Transfer stock to private landlords – RSLs.

Option appraisal

A key part of the strategy is the requirement for the council to develop an Option Strategy.[14] The council should, in consultation with its tenants, review its stock and agree on the most appropriate investment option(s) it will utilise from those set out by the government and outlined above. There is a requirement for this study to be 'signed off' by the government offices for the regions following consultation with all parties including the Construction Housing Task Force.

Arm's length management organisations

An arm's length management organisation (ALMO) is a company set up by a local authority to manage its housing stock. The houses remain in the ownership of the council. Additional funding is available for local authorities to set up ALMOs but the ALMO must achieve what is termed 2- or 3-star rating following review before the funds can be accessed.

ALMOs are normally companies limited by guarantee with council nominees, tenants and independent members of their board. Existing or proposed ALMOs at the time of writing cover about 20% of the stock and conditional funding of £1.4 billion has been allocated. Allocations in excess of £800 million have been proposed for each of the financial years 2004/5 and 2005/6.

Housing PFI

Housing PFI[15] is the award of a long-term contract to a private contractor for it to refurbish properties specified in the scheme and maintain them for the remainder of the contract period, normally 30 years. In a Housing Revenue Account PFI, the houses remain in council ownership but non-Housing Revenue Account schemes are commonly with RSLs involving new build, acquisition or refurbishment with continuing management and maintenance. In the case of the latter, the ownership resides with the operator.

Housing transfer

With the housing transfer option[16] the council transfers some or all of its stock to a housing association. The transfer must be supported by tenants, and where the proposal is for more than 499 units approval by the Secretary of State must be obtained.

This option is not new as it is a model that has been in use since 1988.

Major Repairs Allowance (MRA)

As well as the investment options described above, in 2000 the government announced a major change in the way in which local authority housing would be financed.[17] The Housing Revenue Account (HRA) is a record of expenditure and income relating to an authority's housing stock. In the measures announced, the HRA account is required to reflect the resources used over the lifetime of the asset rather than simply the amount spent each year. This requires councils to determine an annual amount using a discounting methodology based on a new evaluation of the capital value of the stock. The principle is that this should reflect wear and tear of the asset. Within the HRA, therefore, is a new element of subsidy referred to as the Major Repairs Allowance.[18] The MRA represents the capital cost of keeping the stock in its current condition. It is emphasised that the MRA is part of the overall funding received for housing. Importantly, the money can be used for any capital expenditure on HRA assets and councils are free to spend the money outside the financial year to which it relates. Other funding sources are, however, used for day-to-day maintenance. Strict definitions for funding usage are provided in this respect. The notion of keeping the stock in its existing condition is simply a device used to determine funding. The 2002 spending review determined that in 2001/2 £1.6 billion would be transferred from other areas to finance MRA, equating to approximately £560 per dwelling.

The MRA is calculated by establishing on a national level the timings and costs of element replacements for each of 13 classified building types and then determining each authority's MRA by multiplying this national average by the number of dwellings they have in each category. The categories are based on type of construction, configuration, size and age. The EHCS data is used as a basis for determination of the categories described and councils are given detailed guidelines on how to access and interpret this data.

Subsequently, the 2004 Housing Act (Part I) contains provisions to replace the housing fitness regime of the 1985 Act. This takes the form of a Housing Health and Safety Rating System and gives local authorities enforcement options.

This was initially released in July 2000 to replace the Housing Fitness Standard that had been the basis of the 1996 EHCS.

The principle of the system is that:

A dwelling including the structure, the means of access, any associated outbuildings and garden, yard and/or amenity space, should provide a safe and healthy environment for the occupants and any visitor.

To satisfy this principle:

A dwelling should be free from unnecessary and avoidable hazards and where hazards are necessary or unavoidable, they should be made as safe as reasonably possible.

For these purposes a fault is a failure of an element to meet the ideal, whether that failure is inherent, such as a result of the original construction or manufacture, or as a result of deterioration or a want of repair or maintenance. The ideal is the currently perceived model of an element which defines the functions and safest performance criteria that can be expected of that element.

An element is any component or constituent part, facility or amenity of a dwelling, such as a wall, a window, a staircase, a bath, a means of lighting or a means of space heating.

A hazard is the effect that may result from a fault and that has the potential to cause harm.

The system places the emphasis on hazards – the effect rather than the fault – so that effectively the cost or extent of remedial work is irrelevant. The system effectively also implies a risk assessment approach which means that all faults need to be recorded in order to assess whether or not they may constitute a hazard now or in the future.

Within the context of this major overhaul, the 2001 EHCS was the eighth and last five-yearly survey. From that date on the survey will be conducted on a continuous basis.

In terms of the profile of the stock, the 2001 EHCS reported:

❏ There were 21.1 million dwellings, compared to 20.3 million in the 1996 survey.
❏ The stock is relatively old, with circa 40% built before 1945 and 21% before 1919.
❏ Newer dwellings are smaller – the average post-1980 is 83.8 m² compared to 88 m² before 1980.
❏ The North and South-East have the oldest stock and are most likely to be terraced.
❏ 94% have central heating, 76% double glazing and 35% a second WC.
❏ There are 20.5 million households.
❏ Households are generally becoming smaller.
❏ The number of vacant dwellings has fallen from 4% to 3% and these are more likely to be problematic in that they need significant repair to bring them back into use.

The government now refers to what it calls 'Decent Homes' in the survey, and to be classed as decent a property must meet each of the following criteria:

❏ reach the current statutory minimum standard
❏ be in a reasonable state of repair
❏ have reasonably modern facilities and services
❏ provide a reasonable level of thermal comfort.

The key findings of the 2001 survey were:

❏ There had been improvements since 1996, with the number of homes classed as not decent falling from 46% to 33%.
❏ In the social sector the proportion fell from 52% to 38%.
❏ All social groups have benefited, but the poorest households still remain twice as likely to be in non-decent homes.
❏ The most common causes for dwellings to be non-decent are failure to meet thermal comfort criterion (26% of the stock), disrepair (9%), fitness (4%) and modernisation (2%).
❏ A third of the non-decent homes were built before 1919.
❏ High-rise flats have the highest concentration of non-decent stock (58%).
❏ There was less progress since 1996 in improving flats.
❏ Little difference in the prevalence of non-decent homes across regional groups was revealed.
❏ The average cost to make a house decent is £7200, making a total cost of £50 billion.

This latter figure contrasts with the £19 billion needed for renovations and improvements referred to in the 2000 Green Paper. Also in June 2003 another ODPM paper[19] referred to a 'need to spend on maintenance and management in the local authority housing stock for 2001/2' of £5.5 billion'.

Table 2.1 Condition data from ongoing EHCS.

	Are non-decent	Fail thermal comfort	Fail disrepair	Fail fitness
Owner-occupied	29.4%	23.1%	8.0%	3.2%
Private rented	49.4%	40.4%	17.1%	10.9%
Local authority	42.7%	34.1%	8.8%	4.7%
RSLs	27.6%	22.1%	5.0%	3.4%

There is some inconsistency here, but it must be borne in mind that the 2003 paper referred to an annual spend to stop the stock deteriorating. This was part of its announcement of the Major Repair Allowance.

The 2005 ongoing survey published on the ODPM website gave the condition data shown in Table 2.1.

The key findings of the 2003 survey focus on issues of vulnerability, deprivation and liveability as well as decent homes and a summary of these findings is given below:

❏ The number of non-decent homes fell from 33% to 31% of the stock between 2001 and 2003.
❏ The social sector has improved at a faster rate than the private sector.
❏ The most common criterion for non-decent is thermal comfort although improvement in this criterion is the major reason for improvements across the stock since 2001.
❏ Between 2001 and 2003 the proportion of private sector vulnerable households living in decent homes increased by 6%.
❏ Conditions in the most deprived districts continue to improve.
❏ 16% of all households occupy homes with liveability problems related to the quality of the immediate environment. These poor-quality environments are more likely to be in areas with a predominance of deprived households.

Healthcare buildings

Healthcare buildings represent, perhaps, the most difficult group of largely public sector buildings to maintain because of their complex engineering services and their heterogeneous nature. Furthermore, safety and hygiene considerations make the condition of these buildings a particularly sensitive issue.

The situation is further complicated by the radical reorganisation of health services over the last 10 to 15 years, with the formation of health trusts having effectively devolved responsibility for managing their estate. At a national level NHS Estates, an executive arm of the Department of Health, has had until recently the requirement to advise and act as an enabling agency for the provision of an appropriate environment for delivering healthcare. A key component of this is the NHS Estates Performance

Management Division. At the time of writing, however, further change is taking place following an agency framework review. This is leading to a major reconfiguration under which, from September 2005, the majority of the functions carried out by NHS Estates will be delivered via the Department of Health or from within the wider NHS, presumably the hospital trusts.

In April 2002 the structure of the NHS was overhauled with the creation of 28 strategic health authorities and the division of healthcare into primary (GPs, dentists, pharmacists, opticians, district nurses and other close-to-customer services) and secondary (more specialised services provided in fewer locations – hospitals) care. Under the reorganisation of NHS Estates, the Strategic Health Authorities will play a major role.

Another important factor in the health service is the accelerating move towards Public Private Partnerships and PFI contracts in particular. This effectively outsources a great deal of the responsibility for maintenance. The NHS has also been the major public sector division to embrace a holistic approach to FM with bundled or integrated outsourcing.

Since the formation of trusts, accompanied by comprehensive asset/condition surveys or inventories, there has been targeted activity aimed at improving customer care. Much of this activity has been aimed at mechanical services and equipment, as this represents the larger portion of the asset base, and is the most critical in terms of patient care. Nationally, no substantive comprehensive survey data is available on the condition of healthcare buildings. However, in 1986 the total backlog of repair and maintenance in hospitals was estimated to be in the region of £2 billion.

Despite a great deal of activity, amongst maintenance professionals there is a strong feeling that maintenance is underfunded, and that little preventive fabric maintenance is taking place. For example, one maintenance manager estimated that the large hospital for which he was responsible had a maintenance backlog of between £45 million and £55 million. FM had in this case been outsourced but because of a major reorganisation of healthcare a PFI contract was in the process of being let. Part of this would be for a major refurbishment of the largest buildings which would remove much of this backlog under the NHS capital investment strategy. The majority of the current allocation for maintenance was still being used up by planned preventive maintenance on engineering services, much of it driven by the need to meet statutory and insurance requirements.

In 2005 NHS Estates published a strategy for 'Risk Adjusted Backlog Maintenance'[20] in which they issued new technical guidance so as to help trusts take into account more effectively the different levels of risk to patients than previously was the practice.

The existence of a substantial backlog is still widely acknowledged. In 2000 the NHS Plan set a target that by 2010 40% of the total value of the estate should be less than 15 years old (accomplished largely through 100 new hospitals) and that 25% of an estimated £3.1 million maintenance backlog should be cleared by 2004.

The 2002 budget provided for capital expenditure on IT and buildings to be raised over the following two years from £2.2 billion to £5.5 billion per annum with the intention that two-thirds of all hospital real estate and all GP surgeries would be replaced or refurbished. PFI was seen as a major facilitator in this process.

Comprehensive data on the meeting of these targets is difficult to obtain but information drawn from the Construction Products Association[21] and the NHS Estates and Facilities Survey (ERIC)[22] provides useful progress information.

Data shows that 45% of the capital investment is sourced from PFI programmes. The Public Service Agreements accompanying the government's 2002 and 2004 spending reviews did not include specific progress targets but the Department of Health's investment strategy revealed that the government is also seeking:

❑ financial closure by June 2005 on 29 major schemes for new hospitals announced in February 2001
❑ to open 58 additional major schemes by 2008.

Since the publication of the NHS plan the total number of major and medium-sized new schemes has reached 131, which is above target. However, the delivery of new schemes has slipped so that the financial closure prediction for 78 of the schemes will be later than planned. Despite this the number of new schemes approved or on site suggests that the modernisation target will be met by 2010.

With respect to maintenance backlog, the NHS is currently analysing condition data and at the time of writing an up-to-date estimate of the backlog is not available. However, data from the 2002 ERIC survey indicated modest progress in cutting the backlog by 6% to £2.9 billion. The achievement of the NHS target is compromised by slippage in the new building programme delaying the replacement or comprehensive refurbishment of older stock. The Audit Commission has warned from a survey of trusts that:

a majority of trusts were rated as at high risk of failing to significantly reduce the maintenance backlog ... For some trusts, planned large-scale projects will eliminate some of the maintenance backlog, but they will not be completed within the NHS Plan target timeframes.

Educational buildings

There are approximately 25 000 primary and secondary schools managed by local authorities educating over 8 million children in the state education system in England and Wales. Within a statutory framework, local authorities have operational responsibility for school building maintenance, and also have discretion, at local level, for education expenditure and the determination of priorities. They must also, however, determine their overall expenditure within the spending guidelines imposed by central government, which may also indicate the priority it attaches to education through that portion of the annual local authority spending settlement.

There are numerous publications of a non-statutory nature from central government, which as well as providing advice and guidance for local authorities give an indication of central government concerns.

Investment of public money within the education system totals nearly £20 billion a year of which £15 billion is now delegated directly to schools.

Over the last 20 years central government chose to seriously restrict local authority spending by a series of performance penalties. These restrictions caused local authorities to protect essential services at the expense of what were considered less essential priorities including building maintenance. This led to money initially earmarked for operations such as planned and planned preventive maintenance being redirected to other services. Maintenance expenditure has thus been governed by the supply of funds, rather than on the basis of building need.

There has, therefore, been a steady decline in the condition of the education building stock since the early 1980s due to squeezing of resources, but also a number of other factors not least of which was the identification of major condition problems with the system-built stock of the 1950s and 1960s. A number of extensive inherent faults emerged such as deterioration of flat roofs, concrete cladding panels and services.

There was, in the 1990s, growing unease over the condition of this large and diverse group of public sector buildings, and this was the subject of a National Audit Office report published in 1991.[23] Based on November 1987 prices, they estimated a backlog of repair and maintenance of £2 billion within which the cost of structural maintenance alone was estimated at between £730 million and £995 million.

These figures were derived from the requirements of the Education (School Premises) Act of 1981. Under this Act, local authorities were required to bring all existing schools to a defined acceptable standard by 1991. This period was extended to 1996, and a review of the regulations was undertaken in the context of the changes that have taken place in educational practice and management.

In 1985 the then Department for Education and Science (DES) stated that a significant number of pupils were being taught in schools that were in an unsatisfactory condition, and pointed out the detrimental effect this was likely to have on achievement of educational standards. The standards of management used in the control and execution of maintenance work came in for severe criticism. In 1988 and 1990 the National Foundation for Educational research was commissioned by the National Audit Office to carry out research into the problem. They concluded in their 1990 study of 12 local authorities that:

> the difference between what authorities are spending and what they would like to spend on maintenance work remains substantial. Overall, the findings of the survey commissioned by the National Audit Office underscored concern by the local authorities about the condition of school buildings and the backlog of repairs as originally identified by the survey work undertaken by the Department.[24]

Research and discussions with local authorities revealed that priority was given to safety, security and essential repairs, and that routine maintenance and decoration received less attention. This undoubtedly leads to increasing rates of deterioration and accelerates the growth of maintenance backlogs.

An important contemporary issue is the change that has taken place in the management of schools, and in particular the delegation of responsibility for minor maintenance work to the governing bodies of individual schools. The Education Reform Act (1988) introduced the concept of local management of schools (LMS). This gave empowerment to head teachers and governing bodies, and the responsibility for direct management of funds previously administered by the local authority for FM functions such as buildings, grounds and cleaning maintenance. A number of allocation formulae were devised by local authorities but these tended to be based on pupil numbers weighted to take account of school characteristics such as age, condition and location.

Under this framework schools could move funds between budget headings, leading to frequent redirection of funds allocated for maintenance.

Fair Funding: Improving Delegation to Schools was introduced in April 1999 under the Schools Standard Framework Act. Fair funding replaced LMS and provided for 100% management and funding delegation for all structural maintenance to schools in respect of 'concurrent' work. The framework, however, stipulated that funding allocated for maintenance could not be transferred to other budgets. The prioritisation of maintenance funding was still at the discretion of local management. Schools could also take responsibility for commissioning a wide range of minor building works related to maintenance.

In February 2000 the then DFEE issued guidance on two new funding initiatives aimed at addressing the by now well-recognised school maintenance crisis.

Formula Capital funding for schools was an allocation to each school based on the 1999 year 7 pupil numbers. Agreement on what the allocation was to be spent on was subject to mutual agreement between the school and the local education authority although the guidance stipulated that the first year's funding should be based on most pressing needs.

The capital funding allocation would be controlled overall by priority ratings identified within a condition survey produced in line with the requirements of an asset management plan (AMP). The capital funding programme stemming from this initiative is given in Table 2.2.

The second initiative, 'Seed Challenge Capital', offers help to schools that are able to raise their own funds or find innovative ways to lever new money to fund projects

Table 2.2 Education buildings capital funding allocations based on AMPs.

Financial year	Budget allocation
2000/2001	£191 000 000
2001/2002	£260 000 000
2002/2003	£500 000 000
2003/2004	£680 000 000
2004/2005	£627 000 000 (England only)

Table 2.3 Seed Challenge capital funding for education buildings.

Financial year	Budget allocation
2000/2001	£30 000 000
2001/2002	£35 000 000
2002/2003	£60 000 000
2003/2004	£60 000 000
2004/2005	£60 000 000

Table 2.4 New Deal for Schools funding.

Financial year	Budget allocation
2002/2003	£174 000 000
2003/2004	£381 000 000
2004/2005	£580 000 000 (England only)

from external sources. Schools can bid for a top-up to these funds for individual projects. The funding allocations are given in Table 2.3.

Further funding allocations were introduced, commencing in 2002/03, through the New Deal for Schools (NDS) – modernisation funding. These grants are allocated according to pupil numbers in an LEA and according to relative total condition (estimated repairs needed over the following five years, plus improvements needed to improve the suitability of accommodation to deliver the curriculum). The funding commitment is given in Table 2.4.

Other sources of funding are available but these are managed directly by the local authority. The main source comes via the NDS Condition Funding grants, also awarded on the basis of pupil numbers and relative condition.

Condition need is again identified through asset management plans. The old Office of the Deputy Prime Minister extended and projected funds through to 2006 but with NDS condition funding amalgamated under the single budget heading of modernisation/condition funding. This represents part of an additional £4.2 billion capital investment in schools.

Asset management plans

A common issue threaded through all these funding initiatives is the need for a structured, consistent approach to the evaluation of condition.

Initial guidance on asset management plans (AMPs) was issued in 1998 and the first framework issued in 1999. The aims of AMPs are to:

❑ raise standards of educational attainment
❑ provide sustainable and energy-efficient buildings
❑ provide innovative design solutions
❑ increase community use of facilities
❑ maximise value for money
❑ ensure efficient and effective management of new and existing capital assets.

AMPs should set out information and criteria to make spending decisions. When the three elements of Condition, Suitability, Sufficiency are properly evaluated and integrated, the theory is that effective controls will be in place that direct monies in the most efficient way.

The DfES audits each authority's AMP annually. Failing inspections lead to a dialogue aimed at improvement, and high-performing authorities are rewarded with an increased time period between inspections. Provisional unpublished research based on a study of five Midland authorities has produced encouraging results with respect to the comprehensive nature of the AMP approach and the benchmark tool it provides for a process of improvement.

The range of initiatives is certainly comprehensive, but there is still concern about LMS. There is a suggestion also that, whilst the fair funding initiative is generally welcome, disquiet still exists with respect to management and control of maintenance and there is a reluctance to seek professional guidance.

Whilst the current debate is focused on school buildings, it is important to emphasise that these only represent one portion of the building stock of education buildings. The polytechnics and larger higher education institutions became incorporated in 1990 to form higher education corporations (HECs), and in 1992 the majority of them became new universities. Hitherto, these institutions had been under local authority control, and the cost burden of building maintenance fell on LEAs.

Incorporation effectively took them into the private sector, and whilst, on paper, the gifting of the buildings appears to be attractive, the new HECs have inherited considerable liabilities. Many of the polytechnics, for example, grew out of the development and merger of older inner-city educational institutions which were accommodated in ageing buildings. Furthermore, over their lives, these buildings often suffered from a lack of attention being paid to the proper maintenance of their fabric. Additionally, the current expansion in the participation rates for students in higher education is imposing severe pressure on space in these buildings, leading to very intensive patterns of use.

The traditional universities, which have operated on a more independent basis for many years, have developed more effective estate management regimes but, notwithstanding this, there is growing concern about the ability of all universities to meet increasing maintenance needs as their building stock ages.

Execution of maintenance work

The agencies that carry out repair and maintenance work are contractors and, in both private and public sectors, direct labour organisations (DLOs). Figure 2.10 shows the breakdown of the execution of maintenance work between contractors and DLOs. The proportion of work executed by contractors has increased steadily from around 85% at the beginning of the period to 95% in 1988, since when it has remained relatively steady. The figures from DTI also show that the proportion of contractors' output attributable to repair and maintenance has increased from 42% to 47% in the period. This will be due to a number of factors including the influence of CCT and latterly best value, particularly as PFI accelerates. The figures show that the decline in directly executed maintenance work is across the board in this period, that is, in both housing and non-housing.

Although a significant amount of construction is also undertaken by a thriving DIY sector, its contribution to output is not included in government statistics, even though it is considered to be a major contributor to repair and maintenance output.

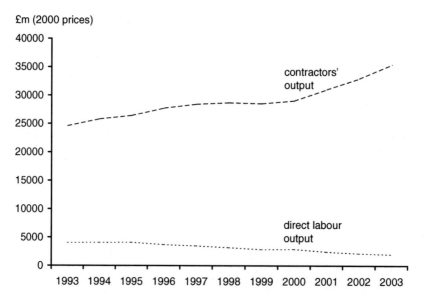

Figure 2.10 Contractor and direct labour repair and maintenance output at 2000 prices. *Source:* DTI

Building maintenance and the professions

A major characteristic of the construction industry is the multiplicity of parties who may be involved in any construction project. There are several reasons for this, but the major contributing factor is the traditional division between design and production, which typifies the industry and leads to a separation of the design-related professions from those more closely associated with production. Franks notes that:

> In no other industry is the responsibility for design so far removed from the responsibility for production.[25]

Whilst it is normal to see the architectural and the contracting professions as epitomising the conventional divisions, there are further divisions within these groupings. Even allowing for contemporary thinking about procurement, which attempts to address traditional divisions, it would be wise to accept that there are differences between architects and builders which are educational, social and cultural, and that these will not disappear overnight.

Another cause of fragmentation, in respect of responsibility for construction, is the range of clients served by the industry. Large client organisations may also employ professional advisors, in addition to the architect, to assist in managing complex projects. In some cases the building owner and the building user, who are not necessarily the same, may both be involved in the procurement process. When the position of building maintenance, within this tangled web, is considered, a complex picture results.

The assessment of repair needs and the execution of maintenance of a building in use brings into play another set of parties who are rarely related to the team that procured and constructed the building, perhaps with the exception of the building owner, who may or may not have had some influence at the design stage.

The role of the building owner with respect to maintenance considerations during design will vary depending on their interest in the use of the building. The assessment and management of maintenance increasingly involves the chartered building surveyor and multi-professional teams particularly through the increasing use of Public Private Partnerships. Best value and the plethora of initiatives in the public sector, particularly some of those described above, are increasingly driving a more integrated and holistic approach.

Despite this and even within such arrangements, the execution of maintenance work itself, which usually makes use of a production team having no connection with the original construction team, is subject to huge variations on an industry-wide basis. The management systems employed, and hence the involvement of the professions, are hardly ever the same from one organisation to another.

Education and training for maintenance

The report of the DoE Committee on Building Maintenance (1972) contained a forthright section concerned with education and training needs. This report was 'revisited' in 1988, and formed the subject of a BMI special report.[26]

The 1972 report identified the following three levels of education for maintenance:

❑   the education of all those who contribute at a professional and managerial level to the building and maintenance processes

❏ education for supervisory personnel who are in direct control of maintenance operations
❏ education and training for those who undertake maintenance work.

Unpublished work by the Polytechnic of Central London (now University of Westminster) concluded that, whilst maintenance was not central to any of the objectives of the existing professional institutions, some indicated an increased awareness of its importance. There was, in the majority of educational programmes for built environment professionals, disappointingly poor coverage of such important topics as the design/maintenance relationship, economic evaluation and maintenance management. The study identified three main roles and eight sub-roles, for which adequate education and/or training was needed. These are summarised below.

(1) Maintenance specialist
 ❏ building fabric
 ❏ building services
(2) General construction practitioner
 ❏ economics
 ❏ design
 ❏ inspection
 ❏ construction
(3) Clients.

This represents the range of parties and functions associated directly or indirectly with the maintenance process, and was used as the basis for the design of an education programme in building maintenance, which was in turn influential on later provision by the Chartered Institute of Building (CIOB).

However, there is little evidence of real movement in providing the dedicated training identified all this time ago. Table 2.5 gives the breakdown from current (2005) CITB figures. Of course this does not tell the full story, as many operatives trained in one of the more popular trades are likely to gravitate into maintenance work ultimately.

Table 2.5 Breakdown of construction training in 2005.

Qualification	Trainees	% of total
Technical	6430	13
Wood trades	14097	29
Bricklayers	8585	17
Painters	3123	6
Plant operatives	4573	9
General operatives	4084	8
Mechanical engineering	3988	8
Maintenance workers	165	0.33
Other	4108	9.67
Total	49153	100

In terms of the design professions, there has been little movement by the Royal Institute of British Architects (RIBA), and there remains grave cause for concern with regard to the industry's performance in terms of the design/maintenance relationship.

The growth of the Building Surveyors Division of the RICS has helped to give increased attention to building performance, although this is often submerged beneath more glamorous activities such as project and facilities management.

Professor Tim Clark[27] comments on what he sees as a lack of bonding between people as the cause of many of the industry's ills. He goes on to say:

> The key to the building up of an appropriate and regularly used network of training around the maintenance task is to obtain the commitment of its users through involving them in its planning. Planning promotes a sense of ownership and with it comes a fuller awareness of group identity, choices and opportunities.

In general experience, the operation of maintenance work, and the way in which it is included in courses of professional study, in many ways epitomises the ideologically fragmented nature of the construction industry. Nowhere is it more evident that the industry lacks a basic sense of cohesion amongst its members than in considering the performance of the building stock. That a more integrative approach to education is desirable is just as true for maintenance management as it is for building design or production in general.

For some years now there has been a general decline in the level of training within the industry as a whole, especially in craft training, which many people would see as having been diluted. However, this may be less of a problem for general maintenance work, where often the requirement may be for a multi-skilled operative, than in the more specialised areas of conservation work.

The real problem for the industry as a whole is one of volume of training and the future skill shortages this will create. Given the unfavourable position of maintenance when competing for scarce resources, its operation may be adversely affected through lack of suitable manpower. The industry has been generally short-sighted in its approach to training, in terms not only of output, but also of content.

In terms of improving performance, the major stimulus should, in theory, come from the client, and proponents of free market pressures would argue that, in the long run, the organisation that is most efficiently organised, with the best-trained workforce will prosper. If this were true, one would expect to see substantially more effort being diverted to training than is at present evident.

Even if this set of problems can be resolved, other issues specific to the training of maintenance operatives remain to be dealt with. It has to be recognised at the outset that the majority of maintenance operatives gravitate into maintenance work, after having spent a number of years in new-build work, so that their experiential base is a general one. It has been acknowledged for some time that the requirements for maintenance operations are best met by multi-skilled operatives, and present training programmes do not produce this person. There is, in other words, little or no specific training for maintenance operatives. Even if such programmes existed, it is extremely unlikely that an industry that struggles to attract talented people for its more glamorous aspects would succeed in doing so for what is often seen as its least attractive. The best that can be expected is that existing single-skill training programmes will pay much more attention to the interrelationship of skills, and attempt to inculcate a greater awareness in trainees of all aspects of the building process.

This inevitably places a quality premium on good supervision, which represents the point at which the professions and the operatives interface, and where many of the deficiencies of these respective groups have to be remedied. It is at this level where, perhaps, the greatest headway can be made, and where training can act not only to improve the skills of a group of people who are amongst the most dedicated in the industry, but also to raise their status by giving them much more explicit recognition.

Maintenance cost trends

Maintenance work is usually more expensive than new work because of the following factors:

❑ It is usually carried out on a small scale, leading to diseconomies of scale.
❑ There is a need to strip out existing work and generally prepare for repairs and replacements.
❑ It frequently has to be carried out in confined or occupied places.
❑ It is very common for the cost of accessing a maintenance item to be several times that of actually carrying out the repair.
❑ The cost of making good and general clearing away is disproportionately high.
❑ It incurs substantial disturbance costs on the operation of the building and perhaps lost production.

The best co-ordinated source for maintenance cost trends is provided by BMI, a company of the RICS, which has replaced the earlier Building Maintenance Cost Information Service (BMCIS). BMI publishes occupancy cost analyses of buildings, based on cost data provided by a range of building owners, in both the private and public sectors.

The methodology used in the production of the occupancy cost indices is described in *BMI Special Report 190*, and they are published regularly in the *BMI Quarterly Cost Briefing*. The indices are derived from published data on construction labour and material costs, and their production requires the determination of appropriate weightings for these inputs as outlined below. *Special Report 190* describes how this has been carried out, and longer-term maintenance trends are evaluated in regular special reports.

Maintenance costs were analysed by BMI, using data from their own occupancy cost studies, to show the breakdown between redecoration, fabric maintenance and services in *Special Report 331*. The breakdown between a selection of buildings for these elements is given in figures 2.11 to 2.15.

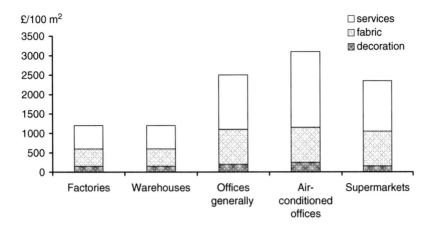

Figure 2.11 Maintenance costs for selection of industrial, retail and commercial buildings. *Source*: BMI

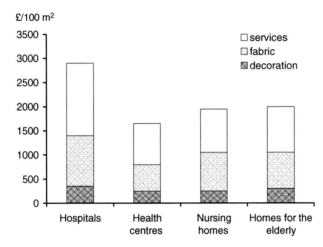

Figure 2.12 Maintenance costs for selection of health buildings. *Source*: BMI

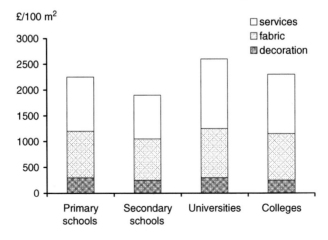

Figure 2.13 Maintenance costs for educational buildings. *Source*: BMI

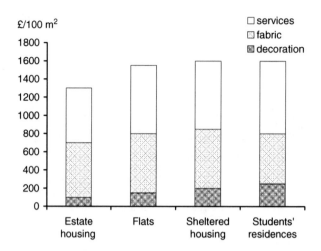

Figure 2.14 Maintenance costs for housing. *Source*: BMI

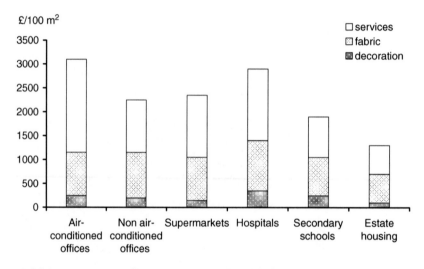

Figure 2.15 Maintenance costs for cross-section of buildings. *Source*: BMI

In *Special Report 331* BMI also carried out an analysis in order to determine what they considered a reasonable annual maintenance expenditure, recognising that historical figures represent an underspend. It comments:

> It is impossible to know if the levels of expenditure have been sufficient to maintain the condition of the buildings. However several of the public sector reports do indicate a significant backlog of maintenance.[28]

To do this they produced an estimate based on 2.5% of the capital cost so as to provide 'a useful benchmark for long term maintenance budgets'.

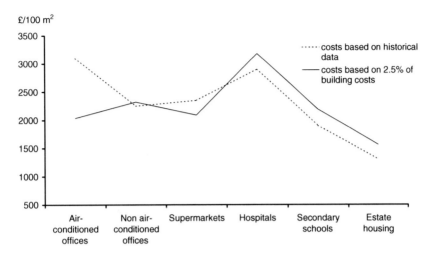

Figure 2.16 Comparison of predicted costs based on historic data with predicted cost based on expenditure of 2.5% of building cost. *Source*: BMI

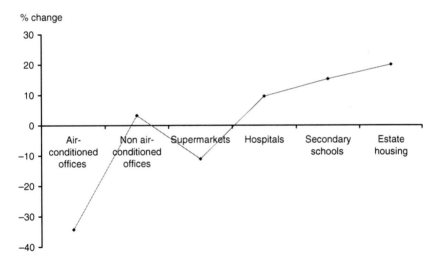

Figure 2.17 Predicted change in expenditure using 2.5% of building costs. *Source*: BMI

Figure 2.16 compares predicted maintenance costs for historically derived figures and those based on the 2.5% of capital cost for a range of buildings taken from *Special Report 331*.

Figure 2.17 shows the percentage change in expenditure that would occur if the 2.5% were used. A varied picture clearly emerges, but there is a clear indication of a possible underspend for the public sector building types.

BMI and the Society of Chief Architects in Local Authorities (SCALA) carried out a further analysis where they compared estimated maintenance costs (based on 2.5% of capital cost) with approved expenditure as a percentage of building costs. The results

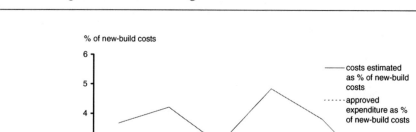

Figure 2.18 Estimated and approved expenditure as percentage of new build costs. *Source*: BMI and SCALA

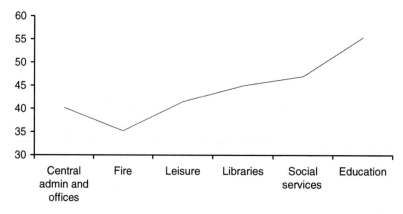

Figure 2.19 Shortfall in funding expressed as percentage of requested funds. *Source*: BMI and SCALA

are shown in figure 2.18. Figure 2.19 shows the predicted shortfall in funding as a percentage of requested funds.

To establish a breakdown of input costs within each of the maintenance elements BMI carried out further work, extending beyond data available from their own occupancy analyses, to give figures for the labour/materials split as set out below:

❑ fabric maintenance   65:35
❑ decoration   89:11
❑ services   65:35.

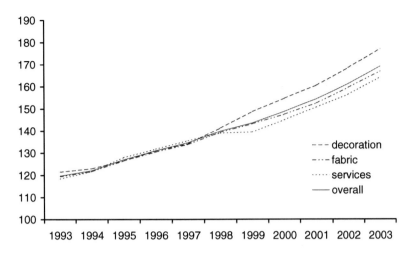

Figure 2.20 Maintenance cost indices – 100 = 1st quarter 1990. *Source*: BMI

For cleaning, BMI accept the extreme variability to be expected in practice, and include for plant in their breakdown. A breakdown between labour, materials and plant of 90:6:4 is used.

Once these breakdowns are established, they are applied to published cost indices. *Special Report 190* lists in detail the cost indices that are used to produce maintenance cost indices using the above weightings. Figure 2.20 indicates current indices.

Special Report 197 gives maintenance cost indices. in compiling the basis for the indices, BMI produced a total occupancy expenditure study for four building types:

❏ administrative and commercial facilities
❏ recreational facilities
❏ educational, scientific information facilities
❏ halls of residence.

This was executed under the following cost headings and an average of all four types of building gave the following breakdown:

❏ utilities   28.0%
❏ overheads   21.5%
❏ administrative   17.0%
❏ cleaning   16.5%
❏ fabric   6.5%
❏ services   6.0%
❏ decorations   4.5%.

In general, utilities (fuel and energy) represent the largest proportion of annual expenditure and fabric, services and cleaning combine to amount to only 17%.

References

(1) British Standards Institute (1984) *BS 3811: 1984 Glossary of Maintenance Management Terms in Terotechnology*. HMSO, London.
(2) Hillebrandt, Patricia (1985) *Economic Theory and the Construction Industry*. Macmillan, London.
(3) Briscoe, Geoffrey (1988) *The Economics of the Construction Industry*. Mitchell, London.
(4) Department of Trade and Industry (2004) *Construction Industry Annual*. The Stationery Office, London.
(5) Hecht, Françoise & Doyle, Neil (1991) Repair cuts will come back to haunt owners, *New Builder*, 25 April.
(6) Building Maintenance Information (1991) Storing up problems for the future, *BMI News* (RICS), September.
(7) Thomas, A. & Archer, P. (1989) Lies, damn lies and the English House Condition Survey, *Environmental Health*, February.
(8) Building Maintenance Information (1992) Maintenance expenditure starts to decline, *BMI News* (RICS), March.
(9) Department of the Environment (1982) *English House Condition Survey 1981*. HMSO, London.
(10) Department of the Environment (1988) *English House Condition Survey 1986*. HMSO, London.
(11) Audit Commission for Local Authorities in England and Wales (1986) *Improving Council House Maintenance*. HMSO, London.
(12) Audit commission for Local Authorities in England and Wales (1986) *Managing the Crisis in Council Housing*. HMSO, London.
(13) DETR (2000) *Quality and Choice – A Decent Home For All – The Housing Green Paper*. Department for Environment, Transport and the Regions, London.
(14) ODPM (2005) *Factsheet 3: Decent Homes Options and Options Appraisals*. Office of the Deputy Prime Minister, London.
(15) ODPM (2005) *Factsheet 3: Decent Homes and Option Appraisals – Housing PFI*. Office of the Deputy Prime Minister, London.
(16) ODPM (2005) *Factsheet 3: Decent Homes and Option Appraisals – Housing Transfer*. Office of the Deputy Prime Minister, London.
(17) DoE (1999) *News Release 602*. Department of the Environment.
(18) DETR (2000) *A New Financial Framework for Local Authority Housing: Guidance on Stock Valuation for Resource Accounting*. Department for Environment, Transport and the Regions, London.
(19) ODPM/BRE (2003) *Estimation of the Need to Spend on Maintenance and Management in the Local Authority*. Office of the Deputy Prime Minister, London.
(20) NHS Estates (2005) *Risk Adjusted Backlog Maintenance*. www.nhsestates.gov.uk
(21) Construction Products Association (2004) *Achievable Targets? 2004*. constprod.org.uk
(22) NHS Estates (ongoing) *Estate Return Information Collection*. www.nhsestates.gov.uk
(23) National Audit Office (1991) *Repair and Maintenance of School Buildings*. HMSO, London.
(24) National Audit Office (1991) *Repair and Maintenance of School Buildings*. HMSO, London.
(25) Franks, J. (1992) *Building Procurement Systems*. Chartered Institute of Building, London.
(26) Building Maintenance Information (1989) *Building Maintenance – Investment for the Future*. Special Report 176/177. RICS, London.
(27) Building Maintenance Information (1989) *Building Maintenance – Investment for the Future*. Special Report 176/177. RICS, London.
(28) Building Maintenance Information (2004) *Review of Maintenance costs*. Special Report 331. RICS, London.

Chapter 3
Maintenance Organisations

The position of a maintenance department within an organisation is dependent on the strategic objectives of that organisation and the importance it attaches to the condition of its buildings. Any study of a maintenance department must consider:

❑ its position within the overall organisational structure
❑ the organisation of the maintenance department itself.

Before either of these can be discussed in any detail it is essential to establish the organisation's attitude to maintenance, as this will influence the policy framework within which maintenance operates. This in turn will have a major effect not only on organisational matters, but also on some other facets, such as operational details, and the approach of the organisation to the procurement of buildings.

The context within which maintenance exists

Buildings must be considered to be a facility and or an asset, and this requires maintenance to be viewed in a wider context. As we have seen, various definitions of facilities management have been proposed. For example, Powell does this in a number of ways:[1]

❑ 'It is concerned with the systematic optimisation of our property and use of our environment.'
❑ 'Facilities Planning determines how an organisation's tangible fixed assets best support achieving the organisation's objectives.'

Powell goes on to comment that FM is also intrinsically linked to change, thus emphasising that buildings management must operate in an essentially dynamic climate. This is becoming an increasingly critical aspect of organisation design, and consideration of the 'contingency' thinking discussed below is very relevant.

Related to FM is the notion of asset management, which tends to focus management attention on those facilities or assets that are perceived as making a direct contribution to corporate profit-making objectives. In many sectors of industry buildings are not viewed in this way, and thus effectively become excluded from active consideration in terms of asset management.

The Audit Commission in their study of the management of local authority assets[2] list amongst their recommendations:

- ❑ build understanding of the strategic importance of property assets including service delivery
- ❑ identify responsibility for strategic asset management.

The role of the facilities manager tends, increasingly, to be involved with those issues perceived by senior management to be important to the efficient usage of buildings, such as space utilisation and energy consumption. Whilst a great deal of progress has been made, there is still too often insufficient active consideration given to the maintenance of the building fabric, often due to poor perceptions of strategic priorities.

Comprehension of the impact that building fabric may have on the productive performance of the occupants of buildings has improved. For example, there is a better understanding of the phenomenon of sick building syndrome, and this is focusing attention on the less obvious effects of poorly maintained fabric and installations.

An apparent lack of interest in fabric condition manifests itself in managerial, technical and financial neglect, which must be considered a facilities management failure. However, there are cases where the buildings are recognised as being more directly related to corporate performance. They may, for example, be a direct generator of income, in which case there is often a presumption that market economics will prevail to force owners to maintain buildings in a proper condition through rents. This assumption of perfect market conditions, however, is rarely accurate[3] and increasingly complex organisational structures cloud these issues further.

Even if it were true, there is no guarantee that this leads to a rational approach to managing building condition. Rather perversely, the response to rent pressures is likely to be a superficial one. The customer in the marketplace tends to be rather unsophisticated in terms of building condition, and often rather easily seduced by a cosmetic response. For this reason serious building condition problems are rarely tackled at the correct time.

Within the public sector a fresh range of problems present themselves. The perceived importance of building fabric varies according to the use of the building, and different standards may be adopted for schools, healthcare buildings and local social housing. However, many of these buildings are not seen as generators of income, but rather as one of the means by which a whole range of society's needs are met. Although market pressures may be limited, there will be other pressures, social and political as well as economic. Almost inevitably there will be a need to reconcile conflicting demands with limited resources, which almost always militates against adequate funding being allocated for building maintenance. Parents of school children, for example, will be much more concerned that a school has adequate book stocks and that there are small class sizes than they will about the condition of a school, so long as it is apparently in reasonable repair and safe.

The best value initiative, though, has begun to focus attention more effectively on the complex range of interdependencies that public sector managers have to cope with. Developments in performance measurement, benchmarking and asset management plans now provide tools through which a much more holistic approach is developing.

What emerges from this brief discussion is that maintenance policy has to be considered not only in the light of a building's function, but also in relation to the user's perception of the building's condition, and its relevance to his primary needs. This latter effect is extremely difficult to quantify and will be discussed in more detail later. However, what is certain is that neither facilities management nor maintenance management can operate without a clear picture of what is being managed and what the functional requirements are.

Management in all areas requires the stating of objectives against which performance is judged. The corollary of this, in relation to buildings management, is that the defining of user needs is an essential prerequisite for the evaluation of building performance. Only then does it become possible to determine what action, if any, is necessary to maintain the building's function, and to provide a benchmark against which performance of maintenance operations can be judged. Such considerations should, of course, encompass not only animate users but also inanimate ones, such as plant, equipment and machinery.

Proper maintenance management, therefore, should not only embrace managing the building in use but also play an important part in its procurement. If this process is carried out correctly then a comprehensive performance model of the building should be constructed at the outset, resulting in a skilfully developed design. At the completion of the construction phase, the building owner and the user, who may not be the same, should be in receipt of a facility together with the necessary instructions for its proper use. It is hoped that one of the benefits of contemporary procurement approaches such as PFI and PPP will be to bring about better continuity in this respect. Indeed, the need to carry out value engineering and consider whole life cost is a stated objective in these arrangements.

Maintenance policy framework

BS 3811 defines maintenance policy as a strategy within which maintenance decisions are made.[4] This may be considered as a set of ground rules for the allocation of resources between the various types of maintenance action that can be taken. Maintenance policy should be considered in the widest possible context throughout all the phases in the life cycle of a building. Furthermore, it needs to be recognised that policy influences on maintenance may not always be direct ones. In other words, it is possible to distinguish clearly between:

❏ policy that is specifically directed toward building maintenance
❏ policy decisions taken with respect to other matters, but which will influence maintenance.

The attitude of, or stance taken by, decision makers during the procurement of a building will have a profound influence, not only on the strategic design but also on the building's subsequent performance. All decisions should be carefully examined, and the possible consequences for the building throughout its economic life considered. In simple terms, the building cycle can be described in the following six stages:

- ❏ brief
- ❏ design
- ❏ procurement
- ❏ construction
- ❏ commissioning
- ❏ operation.

These are all considered in detail later, but at this stage it is necessary to identify their relevance to policymakers, and indicate their possible implications for fabric maintenance.

(1)   Brief
This phase in the building's life involves establishing a performance model for the building, as an essential prerequisite for the proper and effective management of that building, including its maintenance. The model sets a standard against which the performance of the building in use can be measured. The importance the building owner attaches to the setting up of this model is indicative of the attitude he is likely to adopt towards property and its use.

(2)   Design
Building design will be subject to a policy stance at two levels throughout the process. Firstly, a position has to be taken at the conceptual level, in terms of the type of building required to perform the function in question. This may be manifested in a number of ways: for example, the budget allocated to it, the time allocated before occupation is required, a specific statement on maintenance, and the anticipated life of the building. Secondly, the development of the detailed design, which follows, should be a natural consequence of initial policy decisions put into motion at the conceptual level.

(3)   Procurement
The basic requirements for the building, identified at the early stages, will require that a considered view be taken of the most appropriate procurement system to be adopted. This may have repercussions on long-term performance, and hence on fabric maintenance requirements. For example, a need for early occupation may dictate a fast-track approach, which will place constraints on the design, and possibly result in consequences for fabric maintenance in the future. It is important to emphasise this causal link, and to stress that the likely outcome of these policy decisions should be analysed to their logical conclusion. Current trends in procurement are of obvious significance here, but at the time of writing any attempt to analyse their effects would be rather speculative.

(4)   Construction

The outcome of the construction stage, which is conditioned by earlier design activities, may be judged by assessing how well the building meets the client's basic requirements. Furthermore, client satisfaction will also be influenced by the quality control exercised by all parties during site operations. The analysis of building defects suggests that whilst designers and contractors share the responsibility more or less equally for building faults, there are instances where policymakers within the client's organisation must also take some of the blame.

(5)   Commissioning

The culmination of the preceding stages in the procurement process is delivery of the building. In too many cases the way in which this is performed is exceedingly unprofessional, not only in terms of administrative and practical considerations, but also in relation to the information provided to the occupier/owner on the asset he has acquired, often at great cost. The effectiveness of the handover and commissioning phase is a key determinant in the subsequent performance of the building, and improvements are only likely to come about when there is an increased awareness of this link by building owners, which will prompt them to demand a better service.

(6)   Operation

The position adopted by management with respect to the occupation and running of their buildings will be consistently subjected to a range of pressures, including commercial, aesthetic, social, political and economic. The essential issue is not so much that maintenance should be given higher priority, but rather that the need for maintenance is recognised in the first place. If competing demands for scarce funds cannot all be satisfied, any decision not to fully fund maintenance work should only be taken after a carefully considered analysis.

The operational phase is inescapably linked with what has gone before. Once the building is occupied its operation and usage fall under the influence of the organisation's policy with respect to all aspects of its mission.

At the time of occupation there are four possible scenarios, all of which may be within a broader FM framework:

❑   Maintenance of the new building will be carried out within an existing maintenance management framework, which represents a continuation of existing policy.

❑   A maintenance management framework does not exist and must therefore be formulated, in which case there should have been early recognition of this so that it is put into place at an early enough stage to inform all stages of the process.

❑   The new building is so significant, in relation to the existing building stock, that it calls for a rethink of existing procedures and systems, which should ideally have been initiated at conception.

❏   The new building forms part of a PFI arrangement with a maintenance management agency external to the organisation.

The last of these has profound implications for the nature of the maintenance organisation and leads to consideration of organisational matrices discussed below.

Maintenance policy issues

Whatever scenario exists, in considering the operation of maintenance management there are a number of common areas requiring a policy statement.

Resource allocation

The proportion of resources that will be allocated to building maintenance will have to be determined in a competitive environment. These resources may be in terms of finance, staffing (both managerial and operative) and time. Generally, maintenance tends to compete on rather unfavourable terms for all of these, and for finance in particular.

Following the allocation of maintenance resources as a block is the need to decide precisely how these resources are to be distributed. Given the inevitable pressures, this may be carried out in a variety of ways, some of which may have little to do with building performance considerations and be beyond the influence of the maintenance technical staff. The process may be the result of a clearly defined policy or of some mysterious internal process dictated by other characteristics of the organisation.

Here too we need to recognise that the resource allocation may be part of an outsourced operation that will have contractual implications to be considered and may be heavily influenced by a bidding process subjected to a range of market pressures.

Performance requirements

If a logical approach to building performance has been taken from inception, then a detailed performance model may exist. This relates of course not only to technical standards, but also to operational and financial ones, such as response times and budgets. An external bidding process preparatory to entering into an outsourced arrangement will need to address all of these questions, and the importance of properly defined service level agreements cannot be emphasised strongly enough.

Execution of the work

A policy will need to be formulated to indicate how maintenance work is to be executed. This will involve consideration of such factors as:

❏   Who executes the work?
❏   When is it executed?

❑ How is it executed?
❑ How is it supervised and controlled?
❑ What is its relationship with other activities in the organisation?

Administrative activities

Consideration of work execution requires an assessment of the procedures necessary to administer maintenance operations, and this strikes at the heart of maintenance management. The type of maintenance department may or may not be a result of a carefully formulated policy, but will certainly be a reflection of the parent organisation's attitude to the maintenance of buildings. The increasing tendency towards a partnering and/or outsourced approach increases the demands on these administrative activities but more importantly changes their nature away from the operational towards one of monitoring and control.

Position of the maintenance department within the organisation

The position of the maintenance department within the organisation, and its relationships with other departments and functions, may be the single biggest indicator of the degree of importance attached to maintenance by senior management. A carefully integrated maintenance department probably indicates a positive policy stance, where building maintenance has been considered as an important part of the organisational objectives. This is obviously related to overall corporate objectives. However, in too many instances the reverse will be the case, which reflects the low priority given to property maintenance by many organisations.

Clearly, where an organisation has outsourced its maintenance, whether part of a comprehensive FM package or not, the positioning of the responsibility for monitoring and control needs to be rethought and it may well be that there is no option other than to locate it close to strategic management levels.

The discussion so far establishes maintenance in the context of the organisation, and has identified in broad terms how policy may directly and indirectly affect the extent of its influence and the constraints placed on it. A large number of important issues have been touched on, the significance of which must be considered further in the context of a business framework.

The business organisation

Organisation theories

Organisation theory has considered three approaches: the military pattern, the human relations determined and the systems controlled. There is now a wealth of reference material available that espouses principles which, at one time or another, could be applied to maintenance organisations.

Traditionally, it was argued that organisational structures should follow the formulation of strategy, and whilst this is logical, the reality is that they evolve over extended periods of time and therefore have to be adaptive. The classical approach stems from pioneers at the beginning of the twentieth century and is characterised by the work of Taylor and Fayol. What emerged from the classical view of management was a deterministic perception. The principles were held to be universal truths about management. From this, organisation structures emerge that are essentially rigid.

Recognition of informal organisation structures alongside the formal, and the shortcomings of classical organisation theory, saw the emergence of a behavioural approach. This espoused the view that the study of management should be centred on understanding interpersonal relationships. The behavioural responses of people in organisations cannot be expected to conform to predetermined expectations. Many theories of leadership also stem from the work of the theorists who developed these ideas.

The contingency theory of organisation design says that organisation is a function of the task to be carried out and its environment. There is a strong link from this theory to so-called organic patterns of organisation. This contrasts strongly with organisations that can be said to be mechanistic and which are characteristic of classical management theories.

Organic organisations evolve naturally to adapt to environment, and those based on contingency theory are best suited to make this response. They tend to be participative in nature and therefore utilise elements of behaviour theory.

Contingency theory believes that managers have to respond to the environment within which their organisation operates. Strategic contingency theorists believe that, notwithstanding this, managers have freedom of choice, and although the environment may constrain these choices it does not determine them. This approach recognises the role of power as a determinant and, given the dynamic nature of the context within which maintenance work is managed and executed, is of particular relevance.

Each of these theories characterises organisational patterns, management styles and administrative systems associated with them, and although in practice it is unlikely that an organisation will fit absolutely into any one, it will tend to exhibit tendencies. At best, it is only possible to comment broadly on these tendencies, as it is rare for an organisation to be designed from first principles, and the maintenance organisation, in particular, is unlikely to be the result of a great deal of forethought.

Business organisations evolve as a company grows, and will therefore be a function of the rate and nature of this growth. In this respect there is a need to be concerned not only with size in absolute terms, but also with the diversification geographically, and the nature of the output. Many companies begin from very humble origins, where management is in the hands of a small number of people. As the company expands, it becomes necessary for a more formal organisation structure to develop, and for convenience rather than accuracy this is often represented by an organogram (figure 3.1). At this point, notions such as span of control, delegation, accountability and authority have to be taken into account. This may, or may not, result in a formally designed

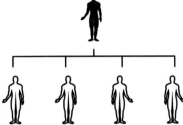

Figure 3.1 Sole proprietor.

organisation structure, with fully defined roles, responsibilities and job specifications, although modern management thinking tends to espouse a more scientific approach to the issue.

There is now a much wider acceptance of management education as a preparation for a career in business, and a much more open-ended debate about the principles, which are increasingly seen as being at the socio-technical-economic interface. Handy, for example, has equated the norms and values that determine the way in which an organisation is operated and structured to cultures.[5] Within this, he considers the following:

❑ *power culture*, which depends on a central power source, typical of small family businesses and characterised by rapid response to change because of the low level of bureaucracy
❑ *role culture*, where roles are precisely defined, which is characteristic of many public sector organisations, and might be considered bureaucratic in nature
❑ *task culture*, in which people are organised in teams around specific tasks
❑ *person culture*, where the individual is the central figure, the best examples of which are specialists who seek to operate independently within an organisation.

On the other hand, some theorists embrace a system view, where the organisation is seen as an adaptive system dependent on measurement and correction through information feedback.

These two extremes are interesting, and it is clear that many of the features of the systems view are an essential part of many modern organisations – a tendency that has been accelerated by the rapid developments in information technology.

Formal organisation structures

The organisation structure may be represented by a chart, or organogram, showing the formal allocation of responsibilities between personnel, and the relationships that should exist between them (figure 3.2). The weaknesses, representationally, of the organogram must be recognised as it is clearly unidirectional in nature and is characteristic of a classical approach.

This may be backed up by a corporate plan, which sets out the general obligations and responsibilities, perhaps accompanied by a policy statement and a clearly defined set of job specifications, or descriptions, for the personnel in the organisation (figure 3.3).

Organograms may be useful, in that they provide an overview of the organisation structure, and may be an aid to clarity of thought. Although many of these structures may be the result not of a planned process, but rather one of evolution, it is true that major companies often undergo reorganisation exercises that focus on basic principles. This will most frequently be part of a response to events in the activities of the business, such as expansion, merger or take-over.

It is fairly obvious that an organisational structure, as well as being historically influenced, is very much a function of the nature of the business, its objectives and the style of management adopted. This may well be a function of policy, but the policy itself will be the product of people.

It may be, for example, that the ethos of a company favours a rigid hierarchical structure, along the lines of the classical or military pattern, and this may, or may not, be related to the nature of their business. On the other hand, it may be that this approach is inappropriate, or that senior management endorses a different attitude, so

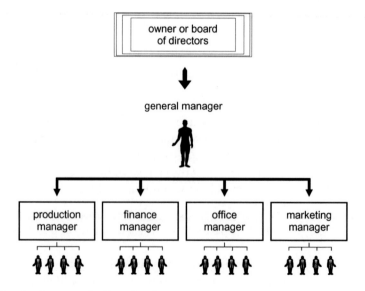

Figure 3.2 Developing organisation structure.

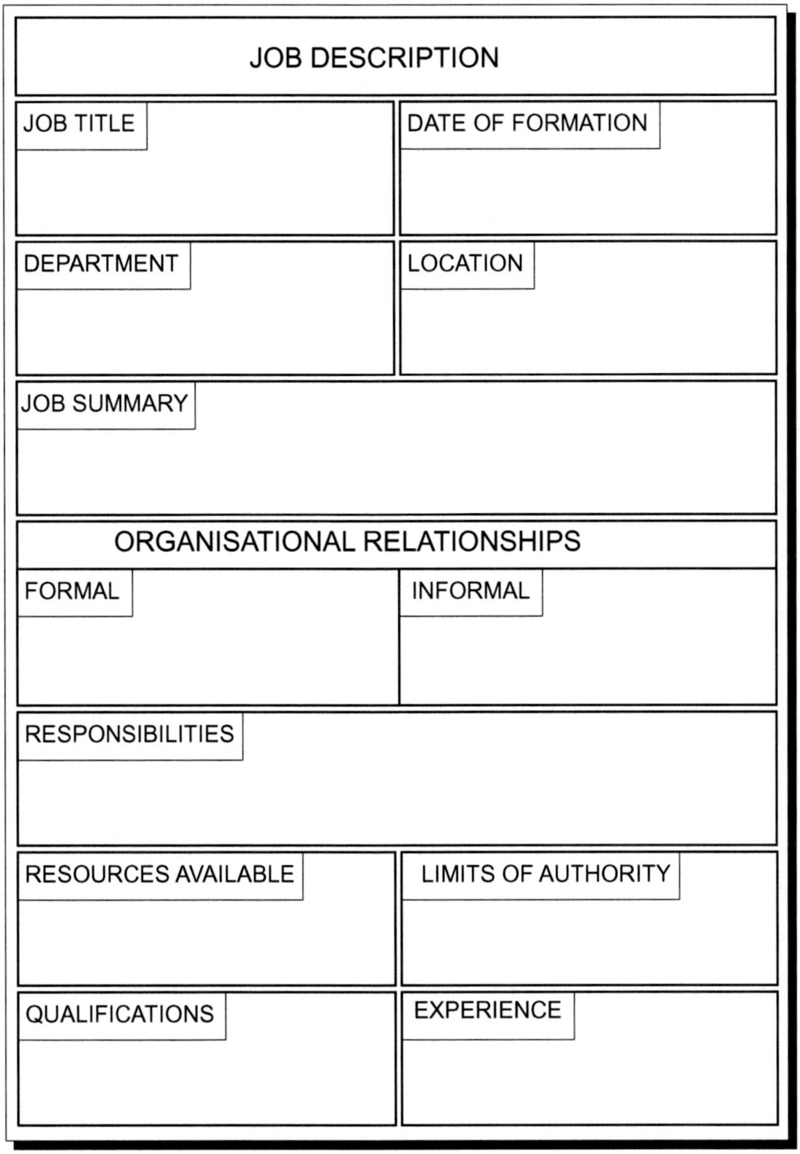

JOB DESCRIPTION	
JOB TITLE	DATE OF FORMATION
DEPARTMENT	LOCATION
JOB SUMMARY	
ORGANISATIONAL RELATIONSHIPS	
FORMAL	INFORMAL
RESPONSIBILITIES	
RESOURCES AVAILABLE	LIMITS OF AUTHORITY
QUALIFICATIONS	EXPERIENCE

Figure 3.3 Typical job description.

that the organisation is designed in such a way as to emphasise the importance of human relationships.

To illustrate the confusion that may arise in passing judgements, it can be said of the organisation of much of the construction industry that it is characterised by a high level of informality, and that this is both its strength and its weakness.

It is also easy to identify a number of companies, operating in similar fields, and of a similar size and distribution, that have very different organisation structures and attitudes to management. The varying approaches that can be taken to construction management, for example, are illustrated in figures 3.4a and 3.4b.

In general, there is little conclusive evidence, in such cases, to imply that one approach or another is the most effective. There are, however, obvious distinctions that may be made between public and private sector organisations, where the context within which each of these works is quite different. Any comparison of performance, under such circumstances, will be difficult. The increasingly complex context of maintenance management, with intricate systems of relationships, compels us to consider more sophisticated thinking.

For some time students of project management have considered the organisation structure in terms of a matrix. This is in fact best illustrated by the scenario where a company outsources FM to an external provider. Maintenance will be part of this arrangement so that the FM provider will subdivide its activities into packages (which may themselves in turn be 'sublet'). In a sense, this can be considered as a project, albeit one with a limited but extended timeline.

The work will effectively be broken down into activities, sometimes represented by a work breakdown structure. Both the client organisation and the external provider will possess an organisational hierarchy (probably on a functional basis), which will probably be different.

The relationship of the work breakdown structure to these organisational structures will probably pose a number of questions, particularly for the client organisation. It will be faced with some hard thinking as to how it deals with these issues within its functional hierarchy. It will almost inevitably have to modify it.

If the organisational thinking is based on conventional fairly rigid classical thinking then there is a failure to recognise reality.

If the organisational structures of the respective bodies are overlapped, a matrix emerges in which individuals in the client organisation may have both 'project' and functional responsibility. Five types of project organisation emerge from this notion.

(1) *Functional hierarchy*, where project tasks are aligned to relevant operational areas whose managers take responsibility for achieving tasks in their area. There is scope for conflict owing to the differing priorities perceived by individual managers.

(2) *Co-ordinated matrix*, where a project controller is appointed with responsibility for co-ordinating tasks but with limited authority over the demands of the functional hierarchy – e.g. allocation of resources.

(3) *Balanced matrix*, where a manager is appointed to oversee the 'project' and responsibility is shared with operational managers. The project manager is responsible for time and cost and the operational managers for scope and quality. The matrix is difficult to hold together as it depends on the relative strengths of project and operational managers. Power struggles emerge, and it may ultimately default, depending on their outcome, to a co-ordinated or secondment type.

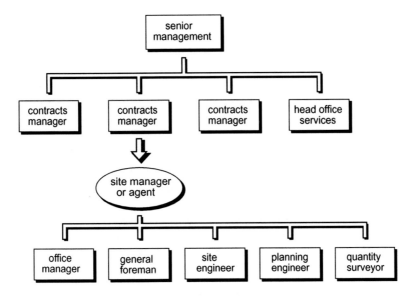

Figure 3.4a Contractor organisation A.

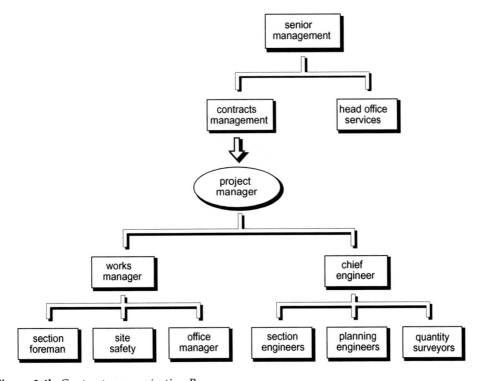

Figure 3.4b Contractor organisation B.

(4) *Secondment matrix,* where a manager has primary responsibility for tasks and the operational managers assign personnel, full- or part-time as required. The project manager has more effective control but potential end-users have lost influence over project outcomes.

(5) *'Project' hierarchy,* where the manager manages a dedicated project team and operational managers have no involvement. The project manager has total control but potential users have lost all influence and the structure is rather inflexible.

Research on project organisations led to the following conclusions:[6]

❏ For *development projects,* which can be considered as those required to carry out strategic changes or developments, for example those that define the purpose of future project activity, user involvement is crucial, and balanced and secondment matrix organisations appear to perform most effectively.

❏ *Implementation projects* are more specific in nature, and a tendency towards more dedicated project hierarchies clearly has most potential.

In considering maintenance of buildings, the latter of these two has more relevance. However, in an integrated procurement scenario such as PFI, we may have to consider both scenarios, and this emphasises the need for flexible thinking.

The matrix style of management worries many theorists who comment that 'much ambiguity must be tolerated and competing claims accommodated for the matrix to function'. Matrix structures are high-demand, high-stress environments and have been cited as over-complex with many quality-control and operational difficulties:

> Almost any organisation will work if the users accept it, but with the matrix system, managers must have clear definitions of authority and responsibility. They must be able to set aside their individualism and understand that the essence of a matrix is that the [project] group must succeed first, then the individual.
>
> The matrix system is a power-sharing, power-balancing organisation. If all the power moves to one side of the matrix, the whole organisation loses. Most decisions are made at lower levels of management in order to push decision-making as far down as possible in the organisation and to encourage participation.[7]

The building maintenance organisation

Scope of the maintenance department

In its broadest sense, the term 'maintenance department' is used to describe the person or persons responsible for the planning, control and execution of maintenance operations. This may be wholly in house or, as is now much more likely, may include independent bodies, such as consultants and contractors. In considering the maintenance management systems to be used, the relationships with these bodies and the rest of the business organisation must therefore be carefully taken into account. The nature of these interfaces will influence operational methods and management systems profoundly.

In general, each of the following phases must be considered in structuring maintenance departments:

- generation of maintenance work
- execution of the work
- control of operations
- provision of feedback
- financial control
- evaluation of performance.

The organisation set up to deal with maintenance needs must address two major concerns. Firstly, it must ensure that an appropriate service is provided within the guidelines established by proper consideration of corporate objectives, and secondly, it must be capable of judging its own effectiveness (or that of an external provider) by monitoring and controlling its performance.

The need to satisfy these two interlinked issues underlines the importance of the interface of maintenance with the rest of the organisation. For example, a typical housing association defines a set of performance indicators, amongst which are the following:

- accountability
- response to repairs
- an indication of the times when properties are unoccupied.

Allied to these, it sets clear performance requirements for maintenance operations, as an integral part of satisfying a set of target performance indicators

The organisation of the maintenance department will normally be determined by the characteristics of the parent organisation and the policy it adopts for its execution, except in cases where fabric maintenance is essential to corporate objectives, for example in a housing association. In this case the maintenance department could have a strong influence in determining the nature of the overall organisation.

Types of maintenance department

It is possible to produce a generalised classification of maintenance organisations according to the degree of domination exerted by one or other of the parties involved in the process. Using this approach, four main parties are identified:

- occupants or tenants
- the owners or client organisation responsible for managing the property
- the 'professional' maintenance team
- the maintenance workforce and their immediate supervision.

These parties are perceived as likely to have different, and perhaps conflicting, goals. Assuming that organisations may sometimes be dominated by one party, four logical types of departmental organisation are thus derived.

(1)   Occupant dominant type
The major emphasis here is on speedy service to occupants, where work is initiated by occupant requests. As quick service is required, the emergency response system becomes overloaded, resulting in highly unpredictable workloads, and probably an inefficient use of resources. Management is concerned with keeping good relations with occupants, and control systems are related to speed of service rather than productivity. Costs are difficult to predict, but tend to be relatively high through the need to maintain large labour strengths capable of meeting peak demands. The labour force will tend to be in house.

(2)   Owner/client dominant type
The major goals here will be:

❏   maintaining the value of the property
❏   keeping costs as low as possible
❏   ensuring that properties are let or utilised as soon as possible.

Work tends, under this regime, to concentrate on external painting and repairs during occupancy, and internal refitting and decorations at changeovers. Management therefore develops a system of maintenance planning related to occupancy patterns, rather than building fabric needs. Frequently, contract labour will be used, owing to extremely variable workloads. Thus management will focus heavily on contract control and competitive pricing.

(3)   Professional dominant type
When this group dominates it is most likely that the style of management will reflect a sympathetic attitude to the maintenance needs of the built fabric. For example, there is likely to be a strong emphasis on planned preventive maintenance programmes, which are seen as limiting the amount of random or emergency work through a proper care regime. If applied rigidly, control will be based on achieving quality, and in some cases may result in unnecessary work being carried out. The workforce will probably be a highly trained one, working within a carefully laid down pattern.

(4)   Workforce dominant type
Maintenance work, in this case, is influenced largely by operative work groups and their immediate supervisors. Standards will be heavily affected by trade group norms, and there will thus be large variations in the quality of work. Control and management will be a grass-roots affair, with a very minimal professional organisation, and thus low overheads.

To this we should perhaps add 'external agency'; however, it is likely that this will exhibit similar characteristics to the 'professional' maintenance team type of organisation, and thus there will be low overheads.

Within each of these, characteristics of Handy's cultural approach are clearly apparent. Whilst not a very effective classification basis, the foregoing categories are useful in terms of what they disclose about maintenance organisations.

Another approach that can be taken is to classify a maintenance department in terms of the characteristics of the building stock and the nature of the organisation, including the importance attributed to maintenance within it. This gives rise to three variables.

(1) The importance attached to fabric maintenance. This can be considered in four categories:

❑ category 1 – primary importance, e.g. housing
❑ category 2 – secondary importance, e.g. commercial with owner-occupier
❑ category 3 – tertiary importance, e.g. commercial rented
❑ category 4 – peripheral importance, e.g. industrial.

(2) The structural characteristics of the organisation
The structural characteristics of organisations are extremely varied. There is clearly a world of difference between servicing the maintenance needs of a large concentrated estate and dealing with a geographically, or divisionally, dispersed one.

(3) Characteristics of the building stock
The characteristics of the building stock can be described by each of the following:

❑ age
❑ type
❑ condition.

In structuring a maintenance department, the degree of homogeneity displayed by the building stock will be a significant factor. Where, for example, an organisation's building stock is heterogeneous, the decision may have to be made to sectionalise the maintenance effort, whereas with a homogeneous stock the need for specialist divisions is unlikely to be necessary.

The two variables relating to characteristics of the organisation and the building stock can be combined to give the following categories:

❑ concentrated homogeneous
❑ concentrated heterogeneous
❑ geographically dispersed homogeneous
❑ geographically dispersed heterogeneous
❑ divisionally diverse.

Functions of a maintenance department

Advisory function

This can be seen as a key area of interface, involving liaison with owners and consultation with senior management, to advise on such matters as:

❏ the development of the brief for new buildings, their design and procurement
❏ the production of as-built drawings and maintenance manuals
❏ the performance requirements of new buildings in general
❏ the provision of specialist advice, and other services related to the areas of adaptation, refurbishment and extensions/modifications
❏ determination of standards to be achieved, and the setting of performance indicators in relation to the primary needs of the organisation, e.g. quality and response times
❏ provision of ongoing information on building condition, which in conjunction with financial information may help senior management in budgeting decisions, and also on decisions as to whether to repair, replace or renew
❏ ongoing information relating to maintenance costs, to assist in sensible financial management
❏ advising senior management on the organisational needs of maintenance, to ensure that an efficient organisation exists, with the correct relationship to the rest of the organisation.

Organisational function

This must be considered with respect to internal functions, and also with points of interface, both within and externally, so that each of the following may be relevant.

(1) The formation of a basic internal administrative system that clearly defines:
  ❏ roles and responsibilities
  ❏ organisational interrelationships
  ❏ communication channels
  ❏ chains of command and patterns of accountability
  ❏ standard procedures.
(2) The defining of proper protocols for dealing with external organisations, and other departments within the organisation. Within this function careful consideration will need to be given to the procedures for communicating information, whether written or oral. Increasingly, information technology is of critical importance when considering administrative and organisational systems.

Operational function

The relevant operations can be classified under the following headings whether they are carried out in house or by an external agency:

❏ identifying the work input
❏ programming the work
❏ ensuring the work is executed
❏ monitoring and controlling quality, cost and time
❏ authorising and arranging payment
❏ providing management information including feedback.

In organisational terms, these operations represent the essential maintenance execution function which, with detailed internal organisational aspects, will be dealt with later.

The typical department structures discussed below are all summarised by the use of an organogram. This formal chart does not always exist, and in many organisations it is informal working relationships that ensure higher levels of service. In other words, there are legitimate situations where the sacrifice of formal control may better serve overall corporate objectives.

Typical maintenance organisations

Local authority property maintenance

The Audit Commission for England and Wales have published a number of reports which present a concise picture of the nature of property management in the local authority sector. The most relevant reports in this respect are two related publications, concerned specifically with management (*Local Authority Property – A Management Handbook*[8] and *Hot Property – getting the best from Local Authority Assets*[9]). These reports advise on wider property management considerations, but provide some intelligence from which useful conclusions can be drawn with respect to maintenance management. They exclude specific reference to housing maintenance, presumably because housing represents a particular set of problems and needs to be considered in relation to the overall provision of what is termed social housing, which is not restricted to local authority properties.

Of the two reports, *Hot Property* is of more current relevance although one of its objectives was to review progress in the light of the many changes that had occurred in local government management. The report identified the following scale of local authority owned property in 2000:

❑ assets valued at around £78 billion
❑ cost of £5 billion a year to run and manage
❑ includes over 21 000 schools, 3800 libraries, 1800 leisure facilities and 2300 community centres.

It divided the property stock into:

(1) property held for the direct provision of services – property occupied by local authority staff for the direct provision of its services, including elderly persons' accommodation, schools and offices, leisure centres and community centres
(2) buildings that support the delivery of services – e.g. council offices
(3) non-operational property – including both vacant property, assets under construction and commercial and industrial buildings.

The mix of properties falling into each of these categories within a given local authority varies with policy and there are also major differences resulting from geographical factors.

The issue of school maintenance is a particularly serious one, but is not given separate treatment here as in organisational terms, and in line with Audit Commission guidelines, it is managed within the wider property portfolio perspective.

The earlier reports identified a series of management problems that all stemmed from what they considered to be an inadequate strategy for managing property. These problems are listed below, and to some extent they may all impinge on maintenance performance. The two that are highlighted in bold, however, are of prime importance.

❑ **inadequate management information**
❑ no incentive to users
❑ failure to carry out property reviews
❑ opportunity cost of holding property not recognised
❑ confused objectives for tenanted/vacant property
❑ **no co-ordinated maintenance policy.**

The latest review concludes that:

❑ Money is wasted on assets that do not support services or are unnecessarily costly to run.
❑ At the same time, key front-line buildings fall into disrepair.
❑ Value for money is unproven – for example, councils own commercial property valued at £7.5 billion but fewer than half of councils regularly review the costs and benefits of retaining this portfolio.
❑ Around 13% of the average civic centre is unused or underused.
❑ Councils are missing opportunities to share property with other agencies.

Earlier reports (in the 1980s) strongly advocated the formulation of a policy statement in the corporate plan and the setting up of a property committee, sub-committee or equivalent body to determine an effective strategy for managing the resource or asset (figures 3.5 and 3.6). The latest report is inherently consistent with this thinking. It strongly advocates the increasing use of more robust asset planning but is far less specific with respect to the advice it provides on organisation structures. Additionally, it encourages a greater emphasis on performance measurement and challenges local authorities to question its practice at every opportunity.

Property management responsibilities should also be clearly defined for members, staff and building occupiers. The reports stated that every authority should have a full property record, and this should include the condition of the buildings. From this a five-year maintenance plan should be prepared, taking account of the age profile of the stock.

Figure 3.7 shows the maintenance management structure within a Shire county council, from which it can be seen that the council has a property department with a director and a number of assistant directors, each responsible for an area of property management. The major interest, in this example, is with the assistant director for property maintenance, whose brief extends somewhat beyond maintenance.

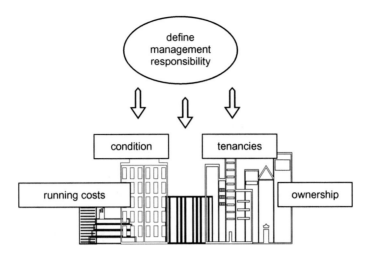

Figure 3.5 Defining responsibilities – an overview.

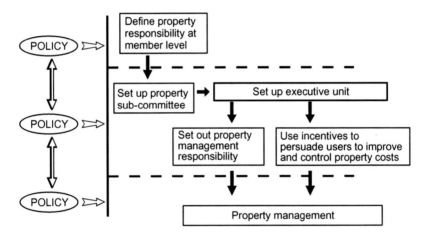

Figure 3.6 Defining responsibilities – local authority.

The 2000 Audit Commission report comments that:

'Ownership' of property by individual committees has been cited as a major barrier to effective asset management . . . operational departments should be held to account for the utilisation of property which is a corporate resource. A dilemma exists: the centre must retain the power to switch property resources between departments but without relieving [the original] department of the primary responsibility for asset management. Authorities **need to establish a trade-off . . . and to spend money wisely.**[10]

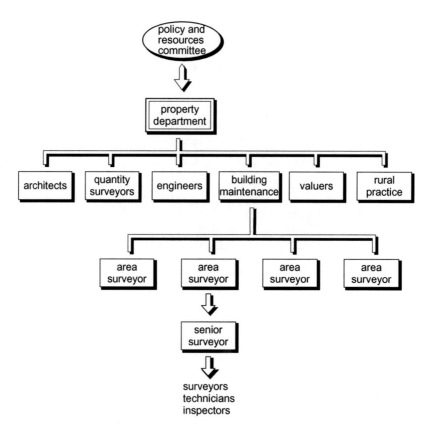

Figure 3.7 Local authority property management.

School buildings

These are worthy of specific mention owing to the manner in which school maintenance is funded as outlined in Chapter 2 and the stock condition issues identified. A key trend in this sector has been for delegation of responsibilities to local school managers. The initial delegation for buildings was limited to non-structural elements and very minor works.

A wider delegation of maintenance funding to schools, known now as devolved funding, came into full effect on 1 April 1999, and represents 100% delegation of structural maintenance to schools in respect of recurrent work. The one stipulation that accompanies this is that funds allocated for maintenance cannot now be transferred to support other budget allocations.

The budget allocations from the local authority to the schools are based on a formula methodology that takes account of school size, age and condition. The prioritisation of maintenance spending is thus in the hands of the local manager but the local education authority remains the landlord.

There are a number of assumed advantages of delegation including:

- ❑   self determination of site (school) priorities
- ❑   the ability to manage maintenance in line with other school activity
- ❑   immediate access to funding
- ❑   flexibility in selection of contractors.

However, there is clearly a disadvantage because of the limited ability of inexperienced, non-cognate local managers to make proper professional building maintenance decisions.

Local authorities recognised this weakness early on and many developed so-called 'buy back' schemes where schools can opt to return their money to the authority for them to manage recurrent maintenance. Schemes varied, but the ability to act as support to schools is only available for money allocated under fair funding and not the powers delegated under the original local management system. This often creates some complex organisational relationships but it has allowed the potential of asset management planning to be exploited more effectively by local authorities.

Housing maintenance organisations

As discussed in Chapter 2, housing maintenance is a major problem throughout all sectors, including the owner-occupied one. Generally, however, the majority of data available represents what we may term social housing with the two major providers being local authorities and housing associations.

Housing maintenance has also, because of its high political profile, received a major share of the attention devoted to maintenance management. As well as its political significance, this sector also has structural characteristics, in particular the homogeneous nature of its stock in terms of building type, and a relatively limited geographical distribution of its stock within a given authority or other managing agency.

Some authorities and agencies have aged housing stock, and many housing associations have a large proportion of refurbished property. Both of these situations cause specific maintenance problems, and may influence the organisation of maintenance operations.

There are a number of other general problems faced by housing management teams in the public sector, including those related to the provision of sufficient funding. This may be a question of total funding, or the way in which allocations are made.

A difficulty with the social housing sector in general is that, although there may be agreement on the objective of keeping the housing stock in a good state of repair, there has been over the years a great deal of confusion as to how this should be defined. It is to be hoped that the 'Decent Homes' initiative and the Housing Health and Safety rating system launched under the 2004 Housing Act (Part 1) will lead to greater consistency to aid performance monitoring. There will inevitably, however, always be different priorities emphasised by tenants, housing managers, maintenance personnel, surveyors, supervisors and operatives.

Tenants, for example, are likely to be concerned with those repairs that seem of most direct relevance to them personally, and naturally want those repairs carried out as soon as possible. The maintenance professional, on the other hand, may prefer a system of planned maintenance that reflects a long-term attitude to maintaining the fabric of the building. The housing manager will be interested in housing as a social service and, when resources are apportioned, the conflicting demands of different types of tenant may be a stronger influence than condition factors.

Social housing has seen many changes in recent years, and constant political pressures are the norm. There have been substantial moves away from local authority control and greatly increased activity in low-cost rented housing from the housing association movement.

The increasing diversity of low-cost rented housing provision highlights interesting organisational factors. It is possible to identify organisations that are owner-, occupier-, professional- or workforce-dominant systems.

One of the major stated objectives of the diversification of provision has been that of breaking down large bureaucratic organisations into smaller units, on the basis that this provides shorter lines of communication – that is, the possibility of a more personalised service. The attractiveness of this was recognised some years ago, and resulted in attempts by large local authorities to decentralise operations, and use area-based work groups as rapid response teams. This can, arguably, be perceived as the marriage of an occupant- and workforce-dominant organisation. The approach has its roots in attempts to counteract perceived social problems and tenant dissatisfaction, to which housing conditions represent one of many contributory factors. This is an example, quite commonly encountered, of housing management taking a stance that may have more to do with social factors than real fabric maintenance.

The accepted wisdom amongst maintenance professionals now, however, is for the introduction of planned maintenance programmes, and organisational structures reflect this.

Over the years a set of recommendations for housing maintenance management have been proposed although there is by no means universal agreement amongst professionals. However, the following principles have been established.

(1) An integrated organisation should exist, with overall responsibility for determining maintenance plans, priorities and customer service standards and for monitoring performance, clearly fixed within the housing department. This recommendation is, however, somewhat in conflict with other arguments supporting a decentralised approach on the grounds that, although costs may rise, service quality improves.

(2) The execution of jobbing repairs should be optimised by careful consideration of the problems faced, and by adopting an organisational strategy to tackle them – for example, through the use of:
  ❏ estate-based repairs
  ❏ zoned maintenance
  ❏ neighbourhood term contracts.

(3) Clear service requirements should be established and performance monitored through an effective management structure.

Housing associations

Housing associations represent a diverse range of non-profit-making private organisations, with a clear mandate to provide low-cost rented housing. The responsibility for their operations is vested in a management committee, and their origins lie in so-called voluntary housing movements. They vary enormously in their size and type of provision. Initially they were intended to operate on a smaller scale than local authority housing departments and to be based in contained geographical regions, operating independently of both local and central government. However, there has been a trend, via mergers and acquisitions, for them to grow in size and become regional in nature and in some cases large national organisations. They are in receipt of substantial public funding through the Housing Corporation which supervises their operations and may also act in an advisory role.

The *1990 Committee Members' Handbook* states that:

> the Committee needs a clear policy on both current and future maintenance and repair. Such a policy must take into account the needs of future tenants as well as those of the present, and under the new Housing Association Grant regime this will include making provision for sinking funds to cover the cost of infrequent and possibly unexpected repairs.[11]

The Housing Corporation is provided with powers under the 1996 Housing Act to issue housing management guidance subject to consultation and approval by the Secretary of State. It introduced in April 2002 a Regulatory Code, together with associated guidance, which sets out the obligations of housing associations. The Housing Act 2004 tidied up a lot of legislation and the Housing Corporation re-issued the Code in August 2005.[12]

The code requires that:

> Housing Associations must be independent . . . and properly constituted not-for-profit organisations. They must operate financially sustainable and efficient businesses and should be committed to and primarily focused upon providing good and responsive housing . . . In doing so they will conduct their business according to the following principles demonstrating their organisations are:

> ❏ viable
> ❏ properly governed
> ❏ properly managed.

These are rather sweeping guidelines, which are supplemented, elsewhere in housing association literature, by further recommendations regarding the need for their operation to:

❑ strike a balance between landlord's and tenant's responsibilities
❑ provide a service that is customer-centred
❑ give good value for money
❑ ensure a consistent operation of policies and procedures
❑ be subject to sound budgetary control.

Housing associations are now rigorously scrutinised by the Audit Commission, which has established a close relationship with them. In their annual reports, most housing associations will include comment on the condition of their stock and some assessment of their maintenance performance. Because of their diversity, and despite clear guidance, there is great variation in performance levels, which does not always appear to be related to their organisational style.

The ethos of the housing association movement has also directed much of its attention to social priorities, such as inner-city programmes, where it enjoyed access to grants for the major repair aspects of rehabilitation programmes.

Figures 3.8a and 3.8b illustrate the organisation of two housing associations. They may both be described as medium-sized organisations within their sector, having between 750 and 1000 housing units.

Association A has a housing stock that is 90% purpose-built, and may loosely be termed new build. Association B, however, has a predominantly refurbished stock. They both operate in similar-sized population centres. Whilst A has a more geographically diverse stock, with a significant number of properties in semi-rural locations, B works almost exclusively in an inner-city environment. These differences, particularly in the nature of their stock, mean that many of the problems they face are rather different.

Association A, as can be seen, has an arm of its organisation specifically dedicated to maintenance. The section is managed by an estates manager, and is responsible for the maintenance function for buildings and infrastructure. There is a well-developed cyclical maintenance programme, with a comprehensive computer-based property file. This database provides an important tool for day-to-day maintenance operations, including emergency repairs. It does not at present relate to major renewal programmes.

Association B has a maintenance department that is under the control of the housing manager, but with important links to the architect and development sections. The operation of maintenance is characterised by a strongly informal set of relationships. Property records are not computerised, but to quote an employee, 'the maintenance manager is a walking encyclopaedia'. There is, however, a strong feeling that major changes will be necessary, if only because of the increasing requirements for housing associations to become more self-sufficient in financial terms. The ad hoc treatment of property records, for example, will not be adequate for future needs. Their major problem is the need for a proper condition survey of their predominantly refurbished stock. Despite this, they do not employ a maintenance surveyor. They do, however, have their own architects and employ some direct labour.

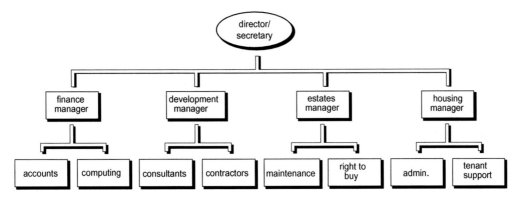

Figure 3.8a Housing association A.

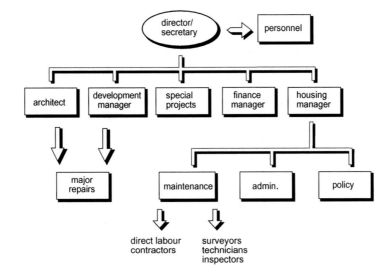

Figure 3.8b Housing association B.

Association A employs only contract labour, and uses consultant architects for all its design services, but it does have a significant maintenance management group, which is singularly lacking in B.

There is here a clear demonstration that the needs of an organisation and its structure are not always matched. It would be tempting to comment that if the organisation charts were reversed there would be a much closer matching of needs and provision.

In both cases, the organisations, as they exist, are the result of historical factors and in particular the influence of personalities.

Comparison of their organisational structures raises an issue with respect to maintenance, which is clearly a central concern of housing associations, and is the subject of much debate in housing management circles. This relates to the question of where the maintenance function should be located in the organisation. Association B considers that, because of its central importance in terms of tenant affairs, it should be an integrated part of housing management. Association A, on the other hand, considers that this approach does not work, as it marginalises maintenance activity. It is, therefore, convinced that a separate maintenance department, with strong management, reporting straight to the director, gives maintenance central importance and status that is clearly on the same level as that of other departments. There is an implicit understanding that all departments will compete for funds but should be able to do so on an equal footing.

Maintenance in the National Health Service

Healthcare buildings generally represent one of the most complex building types in terms of maintenance, owing to their high performance requirements and the complexity of the engineering services needed to sustain proper levels of patient care.

The 2005 use value of the estate is estimated at £23 billion (source: NHS Estates) and at the time of writing this is subject to a massive programme of capital expenditure and modernisation. Like school buildings, it is subject to rapid change in response to its high priority on the political agenda. A consistent theme, however, has been for delivery to be through a range of trusts, self-governing bodies designed to 'localise' delivery and set priorities that respond to need close to the point of demand. The past few years have seen great change in the organisation of these trusts and consequently the manner in which the estate is managed.

The most recent reorganisation has established a structure divided into:

❑ primary care
❑ secondary care.

Primary care is at the heart of the NHS, and Primary Care Trusts control approximately 80% of the NHS budget. Their brief is a local one under which they manage health services at a local level. The key component is general practice, and their remit provides for the commissioning of services (e.g. from Hospital Trusts) to match local demand. They have been in place since April 2002 and report to local Strategic Health Authorities.

Secondary care (often termed 'acute care') usually takes place in an NHS hospital, normally managed as an NHS trust. Secondary care also includes mental health trusts, ambulance trusts and care trusts.

Not surprisingly, along with this there have been a number of changes to the way in which management of NHS Estate is co-ordinated. Up until recently this task was undertaken by NHS Estates, which was formed in 1991 as an executive agency of the Department of Health. The task is now undertaken by the Department of Health Estates and Facilities Division. The organisation of maintenance of health buildings is therefore heavily characterised by a number of influences:

❑ rapid change
❑ a varied range of self-governing bodies

❏   a rapidly changing estate
❏   a dominance of Public Private Partnership arrangements.

Estate management thinking is a reflection of the massive reorganisation of health-care during the 1980s and 1990s. Of all public sector organisations, health is the one most dominated by a more holistic facilities management approach.

Maintenance management within the health service became largely centred around sophisticated computer-based systems at an early stage and this has continued to develop in the form of integrated FM systems often driven from a CAD platform. The production of accurate estate records, including building condition data, was therefore a major priority, but with the continuing modernisation of the estate through the con-struction of new hospitals the quality of estate data has become less problematical.

As in the rest of the public sector, much guidance was aimed at driving estate man-agers towards a much more robust planned maintenance approach. However, given the need to respond to clinical care needs, this is not always easy. Therefore it is not unexpected that much planned maintenance is driven by the need to meet statutory provisions, together with the requirements of insurances, and is dominated by work on plant, equipment and engineering services.

Given the complexities, it is difficult to define a typical organisation structure, and in large part the key organisational aspects will be strongly influenced by linkages to external agencies such as PFI providers. However, figure 3.9 indicates a general outline of an estates directorate for a large NHS trust responsible for several hospitals and figure 3.10 an estates department within this.

The range of responsibility for the directorate as a whole should be noted; this responds to guidance requiring trusts to take a holistic view that integrates capital projects, modernisation, business, hotel services (housekeeping) and nursing services. Facilities projects act as an interface with external agencies. The organisation of the estates department reflects this but the division into response and planned is to be noted, indicating that the thinking is to recognise the different characteristics of these types of maintenance. To manage response maintenance a 'helpdesk' approach is used but a declared objective is to reduce the volume of work to it by actively seeking out work in a proactive way. This is effected by dividing hospitals into zones, visited at least twice a year to carry out as many response tasks as possible. It is assumed that this will also provide a regular inspection approach.

The essential co-ordination of both areas of maintenance activity is the task of the works superintendent.

Fabric maintenance in industrial organisations

(1) Company A
The company whose organisation structure is summarised in figure 3.11 is divided into several divisions in the UK. It is part of a large diverse company, operating on a multi-national basis. It can be seen from the diagram that the fabric maintenance func-tion is under the control of a manager of estate services, who reports to the production

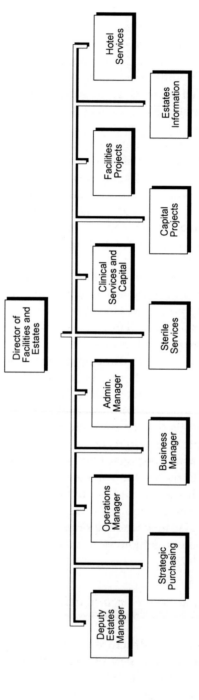

Figure 3.9 Organisation of Estates and Facilities Directorate for NHS trust.

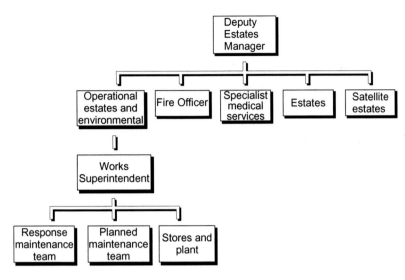

Figure 3.10 Organisation of Estates Department for NHS trust.

manager of the machinery manufacturing division. This division represents the mainstream activity of the company. The estate services section is also responsible for maintenance of buildings and estates for the other divisions.

Geographically, within the UK, the company is centralised in the Midlands, where all the divisions share the same site. There is a strong policy that requires each division to be financially independent. Despite this, there are centralised sales and finance groupings, serving all the manufacturing divisions. Similarly, for operational purposes the fabric maintenance service can be considered centralised.

The organisation chart suggests that the company affords this maintenance function low status, by placing it under the control of one of the divisional managers. In practice, there are also complex lateral and highly informal chains of communication. It is debatable whether this can be effective, and the maintenance section feel very strongly that there is a lack of integration of their function.

There is no planned or preventive maintenance whatsoever, and the section responds to individual requests. This effectively eats up most of the budget for building maintenance. The head of the section clearly considers that they are underfunded, and there is ample evidence to support his view. In terms of corporate objectives fabric condition has low priority in comparison to production.

The section is responsible for building fabric, site infrastructure and building services, maintenance requests being logged into a computer and given a priority rating on a numerical scale from 0 to 9. There are no systematised rules for allocating priority, which appears to be largely determined by who makes the request, although there is a clear understanding that items that directly relate to production should have high priority. Most of these items are building services related – power in particular. This is not surprising in terms of the company's corporate objectives.

Because of the importance attached to keeping production running, each of the manufacturing divisions has its own dedicated maintenance engineer, who is not responsible to the manager of estates services.

Another high priority item relates to building condition at the customer interface. Repairs required to finishes, for example, in areas where prospective clients are received are treated with some urgency. One consequence of this is that much of the section's limited budget is spent on items that the manager considers to be cosmetic.

As is typical in this type of organisation, the individual responsible for maintenance is extremely committed and constantly bombards senior management with requests for funds he considers necessary to achieve what is, for him, a minimum standard.

Maintenance is, in general, carried out by in-house staff, although outside specialists often have to be employed for difficult items. In the case of some large items, company policy is that tenders should be obtained.

From figure 3.11 it can be seen that management rests mainly in the hands of the manager and his general foreman.

Only time will tell if this policy is short-sighted. What is certain, however, is that a time will come, quite soon, when the senior management will be faced with some very awkward decisions, and may well be forced into a major investment of funds. The amount required is obviously growing at an increasing rate, and what may be more problematical is the extensive disruption that will ultimately be caused to their main corporate objectives. There has already been, quite recently, a complete shut-down of a major manufacturing facility, as the result of an electrical failure that could clearly have been avoided by routine maintenance, if the section had had the resources to carry it out.

(2)   Company B

Company B is a constituent company of a major multi-national, high-technology engineering concern, based in the East Midlands. It occupies a site with an area of approximately 25 hectares, containing more than 80 buildings, with a total floor area of $100\,000\,m^2$, dating from the 1940s. The organisational structure is shown in figure 3.12.

Unlike in the previous case, the estates manager here feels that his function is given relative importance. As can be seen from figure 3.12, he reports directly to the assistant managing director, has complete responsibility for maintenance including engineering, and has a staff of over 40. It will also be seen that he is responsible for capital works.

His work falls under four main headings and he budgets accordingly.

(1)   Minor maintenance

This is work he defines as having a low level of material usage, and which is carried out on demand by his own labour. There has, for some time, been in existence a computerised record-keeping system, based on accounting procedures, from which he is able to formulate an annual budget for this area of work. These records also give him enough information to predict the distribution of expenditure throughout the months

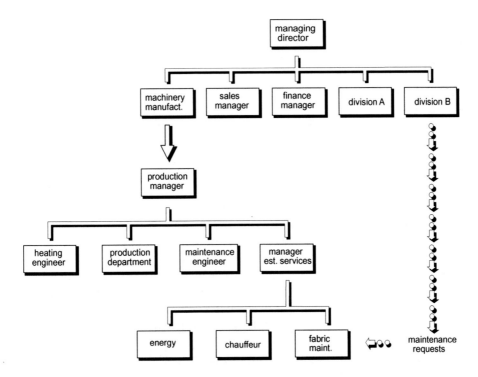

Figure 3.11 Company A.

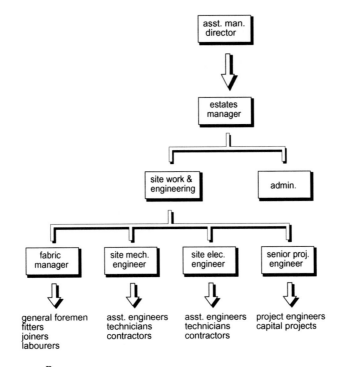

Figure 3.12 Company B.

of the year, and thus allow him a measure of financial control. The current expenditure under this heading runs at around £50000 per annum.

(2)   Planned maintenance

Under this heading he organises the execution of a planned programme of routine work and preventive inspection and maintenance. This is driven, to some extent, by the need to comply with statutory and insurance provisions. It gives a programme of work executed on a range of time cycles, much of it related to engineering services and plant. Other cyclical work is identified from maintenance histories and previous routine inspections. The work is executed by both direct labour and contract maintenance, the latter being on the basis of a schedule of rates contracts. Budgeting is reasonably predictable, given a well-established record-keeping system.

Planning is also well established, through past records, and a known need to execute certain tasks at predetermined times in the year. As an example, electrical substations are always given a regular inspection at Easter, to minimise disruption to production.

Under this heading are also a series of larger one-off programmes, identified from time to time, that have less to do with deterioration than with the changing nature of the company's activities. An example of these is a complete inspection and overhaul of the site electrical systems, the original design of which was based on operations that are no longer carried out. The company operates in a rapidly changing, and highly competitive, international climate, and an important part of the estates manager's role is to try and respond to it.

(3)   Major maintenance

Under this heading fall any other items of a large-scale nature. Whilst the estates manager's funding requests under the previous headings are, in general, met in full, this area of work often represents things he would like to do, but that are not all supported. He receives a fixed budget for all sections of his work but in this third category all the items are given a priority rating from 'essential' to 'if affordable'. When unforeseen emergency items occur, low-priority items are dropped from this budget. It thus provides essential flexibility for the management of his total maintenance budget.

The content of this section of his work is continually under review, and he has sought, where possible, to move what he considers to be very important items to routine maintenance. For example, painting had in previous years been part of the major maintenance section, but for the next and subsequent financial years is being switched to routine maintenance. He justified this to his managing director on the grounds of more effective planning, and economies of scale efficiency improvements.

The total budget for which the department is responsible under these three headings is therefore of the order of £2 million, about 75% of which can be classed as maintenance.

It is worth commenting, however, that the head of this department considers maintenance in a far wider context, and sees all his work as being related to maintaining the site in the optimum condition to help meet corporate objectives, and that refurbishment, conversion work and other capital projects are an integral part of this. Whilst, on face value, maintenance budgets are fixed, the largest element in their make-up falls under the major works section, which is considered to be completely flexible. This provides the potential for response to changing conditions, which is necessary in all maintenance work for this type of industrial organisation.

(4)   Major/capital works

This is the fourth area of work that falls under the control of the estates manager, and the annual expenditure is, as is to be expected, variable, normally in the range from £500 000 to £750 000. It is interesting to note, however, that the estates manager is seen as a major generator of proposed capital projects from his overall role as estates co-ordinator, and the position of comparative strength from which he appears to work.

The organisation is characterised by extremely supportive senior management, and an excellent personal working relationship between the estates manger and the assistant managing director. There is also an extremely effective relationship between the estates department and the accounts department. The latter operates a computerised cost-accounting system, within which maintenance is fully integrated. The system was designed in such a way that constructive maintenance data could be extracted. The essential components of the system were a logical system of numbering all orders, whether for direct labour or contract execution, and the recording of this number on all time sheets and invoices. The secret to the successful operation of the system appears to be the ordering system.

The data generated provides a historical basis for budget-fixing, which means that there exists a tendency for maintenance programmes to be based on past expenditure rather than on need. There are property records stored in a database format, and record drawings which are CAD based. There appear to be no systematic attempts to carry out full condition surveys.

Despite this, the head of department feels that he maintains a good up-to-date picture of the condition of his estate, and is able to respond to building needs through the bids he enters in his major works programme.

Some use is being made of CAD systems for space utilisation exercises but not for maintenance management purposes, although this development is now in the active planning stage.

References

(1)   Powell, C. (1991) *Facilities Management*. CIOB Technical Information Service Paper 134. Chartered Institute of Building, London.
(2)   Audit Commission (2000) *Hot Property – Getting the Best From Local Authority Assets*. Audit Commission, London.
(3)   Spedding, A. (ed.) (1987) *Building Maintenance and Economics – Transactions of the Research and Development Conference on the Management and Economics of Maintenance of Built Assets. Profits and Maintenance Decisions of Owners who Rent for Income*. E. and F. Spon, London.

(4)   British Standards Institute (1984) *BS 381: 1984 Glossary of Maintenance Management Terms in Terotechnology*. HMSO, London.

(5)   Handy, Charles B. (1993) *Understanding Organisations*. Penguin, London.

(6)   Gobeli, D. H. & Larson, E. W. (1987) Project Structures versus Project Performance, *Proceedings of the 11th INTERNET International Expert Seminar* (ed. H. Schelle).

(7)   Poirot, J. W. (1991) Organising for quality: matrix organisation. *Journal of Management in Engineering*, 7, 178–86.

(8)   Audit Commission for Local Authorities in England and Wales (1988) *Local Authority Property – A Management Handbook*. HMSO, London.

(9)   Audit Commission (2000) *Hot property – Getting the Best from Local Authority Assets*. Audit Commission, London.

(10) Audit Commission (2000) *Hot property – Getting the Best from Local Authority Assets*. Audit Commission, London.

(11) National Federation of Housing Associations (1990) *Committee Members' Handbook*. National Federation of Housing Associations, London.

(12) Housing Corporation (2005) *The Regulatory Code*. Housing Corporation, London.

Chapter 4
The Design/Maintenance Relationship

Introduction

Within the framework of the definition of maintenance policy in BS 3811 is the term 'acceptable condition'. The condition of a building is central to the notion of building performance and must be considered throughout all phases of the building life cycle as it is a key element in the formulation of maintenance policy.

A prerequisite for a considered approach to managing building performance is for the organisation not only to have in its possession an accurate picture of its estate, but for it also to take an active role in its development. This begs a number of questions in relation to the manner in which the organisation procures its buildings and, in particular, how they are delivered.

For example, it might well be concluded from a cursory study of many business organisations that there is little apparent relationship between many of them and the design process.

> Maintenance and design are frequently treated as if the two activities were unconnected . . . Maintenance sections often appear to be self-contained . . . [leading to] risk of undesirable divorce from other related functions.
>
> The status of the section and its personnel may be such that there is a lack of influence at policy and strategic levels.[1]

This view compels consideration of the design/maintenance relationship in rather wider terms than at first seems necessary. A design may be executed perfectly well within the terms of reference that have been laid down, but fail to perform properly if these parameters are imperfectly set. It should also be appreciated that, even if a perfect design is executed from an ideal brief, this design has to be developed and realised effectively. Most importantly, the process should culminate in proper delivery of the building to the owner/occupier, together with the information necessary to ensure that it is operated and maintained correctly.

This chapter is, therefore, concerned not only with design in the narrower sense, but also with the whole range of activities from inception to delivery that will ultimately influence the building's performance and its being sustained at an appropriate level.

Performance requirements

It is a requirement in managing anything effectively that there should be a clear set of objectives. In the case of managing buildings, these include the need to maintain the building in such a condition that it is capable of performing the function for which it was conceived.[2] Ideally, this should be represented by a stated set of criteria, against which achievement of this objective can be evaluated. However, for a substantial majority of buildings this statement is not available. The following scenarios are typical examples.

(1) The building is old and has undergone radical changes of use during its life. If the change of use has been the result of a properly conceived refurbishment exercise it should, however, be possible to produce a definitive performance statement.
(2) The building has been conceived in such a way that these criteria were never properly established.
(3) The information is available but has been communicated imperfectly, or not at all, to the client or occupant.

As well as being of concern with respect to the design process, this deficiency will seriously impede attempts to manage a building effectively. Even where there is a clear statement of building performance objectives at the outset, and a perfect solution generated, the situation is a dynamic one, and proper management of the facility must recognise this and respond to it. This requires a process of ongoing monitoring to facilitate effective action. Such monitoring can only be carried out properly by reference to a clearly stated performance model.

All the building systems should be included in such a model, as they will all make a contribution to the effective performance of the building's function. Of course, management judgement and policy may attach different levels of priority to each element, but at least management will be operating within a comprehensive framework.

The process begins at the inception stage and, assuming that the need for a building (or perhaps a major refurbishment) has been identified, the next step is for the client to provide a brief of their requirements for the building. It may be presumed, initially, that a full brief development will address all the relevant issues, resulting in a design that is a proper response to the requirements of the building owner, user and contents.

In ideal circumstances, following this, the design would represent a perfect model of the proposed facility which, as well as being of relevance to detailed design and construction, will be of value to the maintenance manager and the users/owner of the building, thus emphasising the need to view the building as a facility.

The design brief

Having identified the significance of brief development and assumed, perhaps rather optimistically, that this might be perfect and elicit a correspondingly ideal response from the designer, it is now necessary to examine the ways in which this may be properly developed.

Between the initial desire for space, its first occupation and continuing operation, a number of activities have to be successfully undertaken, by a diverse range of people, who may only have an indirect relationship to the primary objectives of the organisation, and an imperfect understanding of them.

The effectiveness with which these activities are carried out will have a major influence on the success or otherwise of the venture. It is, therefore, essential that all parties share a common basis against which these activities can be initiated, planned, monitored, judged and controlled, and that there is a communication system that properly co-ordinates the activities of the people involved.

In seeking the evolution of a solution to satisfy the requirements for a building, four guiding principles can be identified:

❏   to produce a building that is appropriate and efficient for the functions it houses
❏   to produce a building that provides the optimum physical and psychological environment for the contents of the building, both animate and inanimate
❏   to produce a building that strikes an appropriate balance between initial and operating costs
❏   to produce a building that is consistent with the needs and aspirations of the community at large.

In simple terms, the task is to seek a satisfactory design, which is a solution to a set of problems posed by a series of questions. It is the framing of these questions that may be the most difficult part of the process.

Initially a building will be conceived in very general terms, with a broad statement of the building's function – for example, as a factory, housing, school or office – which will then be fleshed out to produce information concerning size, spatial requirements, number of people and facilities. This process often progresses with indecent haste and insufficient real analysis.[3] Even at this very early stage it is essential that the major components and systems of the building are clearly identified, so as to define the way in which they need to respond to the broad objectives. An organised methodology is clearly necessary, and a variety of guidelines may be followed to ensure that the approach taken is a systematic one.

However, simple checklists in themselves are not enough, and it must be appreciated that this should be an analytical process, with proper attention focused on the interface between the building's operational and occupational requirements, and its design requirements in technical and organisational terms.

Although the contents of a building may be both animate and inanimate, it can be assumed that the need for the building is generated directly or indirectly by human needs. The nature of these needs will be derived from the objectives of the person or persons requiring the building, which may be, for example, social, political, economic, moral or religious. The building is, therefore, required to satisfy a set of goals leading towards these longer-term objectives.

Because the process is concerned with human activity, it must be supported by an understanding of the way in which humans interact with their environment. An analysis of user requirements requires consideration of the physiological, psychological and socio-political-economic environment. Human beings are not only shaped by this environment but impinge on it; in other words, they are not passive, and this may complicate the problem.

The provision of any building must also satisfy objectives that are common to a range of potential users, and some that are specific to individuals or user groups, all within an organisational entity (figure 4.1). To create an optimum solution therefore requires the resolution of conflict, and necessitates some compromise. For example, the objective of survival is fundamental. The goals that each individual may see as a means to achieving it will, however, differ as other objectives may interact with this one.

Each person will have a set of goals but no two people will normally have quite the same set (figure 4.2). In the case of buildings, it is necessary to consider not only individuals but groups of individuals that exist as a corporate entity of one form or another. Their objectives and goals will be the product of a complex process which will have taken place in the wider 'universe' (figure 4.3).

To determine the requirements of the future user of a building it is essential to understand these goals and objectives as fully as possible, and this must be the starting

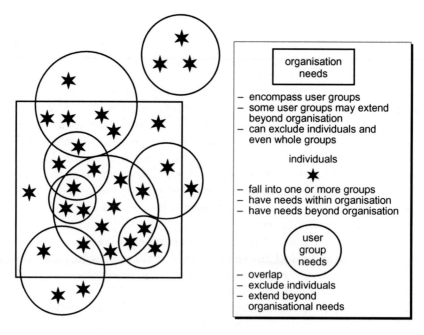

Figure 4.1 Individual v. group v. organisational needs.

	user needs to be satisfied					
	1	2	3	4	5	6
user one	★	★	★			★
user two		★	★	★		★
user three	★	★	★	★	★	★
user four	★	★	★	★		
user five		★		★		
user six		★	★	★		
user seven	★	★	★			
user eight			★		★	
	4/8	7/8	7/8	5/8	2/8	3/8

Figure 4.2 Simple ranking analysis of user needs.

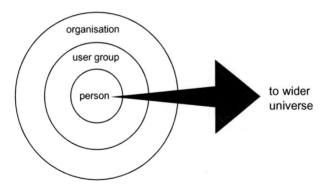

Figure 4.3 Human needs interrelationship.

point for the execution of a user requirement study. The level of understanding developed through this study will inform later studies, and be instrumental in the success or otherwise of the brief.

The next step is to translate what may often appear to be rather abstract ideas into a set of more concrete criteria. The more effectively the client is understood by the designer, then the more accurately this task may be performed.

In this connection a great deal of excellent work has being carried out by the Department of Health, which has done much to address the complexities of designing buildings in the health sector. The Activity Database (ADB)[4] is the Department of Health's preferred briefing and design system. It incorporates an extensive library of room data sheets, layouts and schedules based around schedules of accommodation derived from their health guidance programme. CAD based, it incorporates text and graphic information in a fully exportable format.

Also of interest is the dqi[5] (Design Quality Indicator) toolkit. This tool is an aid to procurement teams to evaluate design quality at key stages in the design process, including briefing. Its development was led by the Construction Industry Council (CIC) with sponsorship from the Department of Trade and Industry (DTI), CABE, Constructing Excellence and the Strategic Forum for Construction. The tool has four versions, one of which is the Brief. The key focus of its operation is to bring together all parties and establish firmly the key objectives of a project. It permits the clear definition of priorities through the application of weightings to key attributes that address functionality, build quality and impact. The tool is designed to be dynamic in that it permits weightings (hence priorities) to be shifted as the project evolves. Used with the Gateway approach described below, it provides a powerful tool for managing the process.

A building performance model

The Building Performance Research Unit at the University of Strathclyde have produced a conceptual model of the building performance system and people, which is summarised in figure 4.4.[6] This represents an idealised conceptual model of building performance (in 1977) and, despite the passage of time, still provides a working model that may be of value in considering the nature of a design brief.

Using this model, the brief can be seen as the analysis of the objective system through which broad goals may be set. This requires reference to the other systems, to help

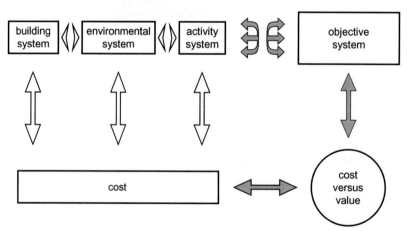

Figure 4.4 Building performance model (after Strathclyde University BPRU).

shape the questions that have to be asked in order to address these goals. Thus, a question and answer routine emerges that facilitates a move from the general to the particular (figure 4.5). For example, the process may start from a broad goal that requires the provision of essential climatic modification. This general requirement may need to be qualified in view of the need to match other aspirations of the client. He may, for instance, have a specific need to meet a certain standard of acoustic performance because of the nature of the business. This may be satisfied by manipulation of either the building system, the environmental system or the activity system, or a combination of them (figure 4.6).

Another climatic modification goal, however, may be a requirement for a building with a rapid thermal response, therefore suggesting the use of a lightweight building system. This will militate against the satisfying of the acoustic requirement through the choice of building fabric, and suggest the need for an alternative solution to this objective, perhaps through an analysis of the activity system.

It is clear that the formulation of a brief through a user requirement study requires the posing of a series of hierarchically structured questions, and that this is an iterative process. The objective of such an approach is to analyse the total building problem, and a properly considered design decision cannot evolve without it. Asking the right questions is only part of the solution, but can increase the expectation of eliciting the information necessary to produce an accurate brief.

A number of diagrammatic techniques are available, but these are essentially informative ways of presenting the problem rather than providing solutions (figure 4.7). In the past some use has also been made of optimisation techniques from operational research. However, the problem is fundamentally a behavioural one.

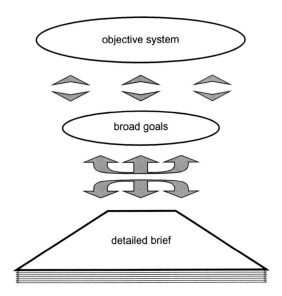

Figure 4.5 From objective system to brief.

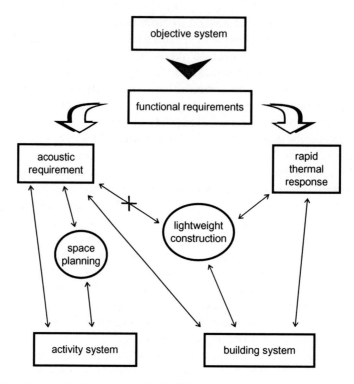

Figure 4.6 Interactive analysis based on the building performance model.

Format of the design brief

Although the analytical process outlined above is essentially an intellectual activity, there is a fundamental requirement for a methodically structured representation of the brief. This will not only be necessary for design purposes, in order to provide a clear statement of the requirements of the building, but will also have utility later for all those concerned with the effective management of the built asset. The format outlined below is one approach that may be taken.

(1) Introduction
  ❑ scope of document and summary
  ❑ contents
(2) Requirements
  ❑ general
    ○ geography and location
    ○ broad function(s)
    ○ approximate size
    ○ timescale constraints
  ❑ occupation density and spatial requirements

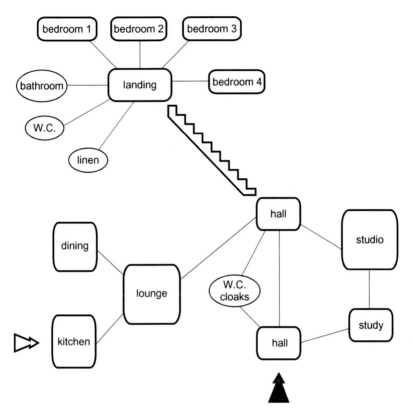

Figure 4.7 Interaction diagram for design of a designer's house and studio.

- ❏ detailed functions, relationships between them and circulation patterns
- ❏ external requirements, e.g. car parking and landscaping
- ❏ communication systems perhaps related to circulation and functional relationships
- ❏ ancillary functions, e.g. catering, cleaning
- ❏ special requirements identified immediately, e.g. air conditioning and IT
- (3) Overall project objective(s)
- (4) Design criteria
  - ❏ space and environment
  - ❏ site works
  - ❏ building shell
  - ❏ interiors and finishes
  - ❏ electrical services
  - ❏ heating, ventilating and air conditioning
  - ❏ plumbing
  - ❏ process and/or production systems
  - ❏ communication systems

❏   fire protection and means of escape
❏   transportation and circulation systems
(5)   Cost constraints/objectives
(6)   Management of the project

The design criteria identified will need to be supplemented by more detailed information, perhaps in the form of a design criteria sheet. It is recommended that these detailed analyses are included in an appendix, cross-referenced to the above items. The design criteria sheets set out the performance requirements of the space, and the activities to be accommodated within the building. A simple illustrative example is shown in figure 4.8.

Maintenance and the brief

Determining the requirement of a building from a study of the user will set in train a decision-making process, which will inescapably impinge on the maintenance requirements of the building. In the first instance there may be a direct effect, in that the broad goals of the building may include a particular stance by the user with respect to running costs. Subsets of goals may then be developed to cover issues such as the minimisation of energy usage and low maintenance characteristics in the building fabric. In such cases, it follows that these items would be overtly identified in a brief statement, such as that outlined above. However, even if low maintenance is not directly on the agenda, a less than accurate definition of the requirements of the building may lead to inappropriate planning and choice of components, materials, and so forth, any of which will lead to poor building performance, with the implications this has for its maintenance.

As long ago as 1978, a NEDO report concerned with client perceptions, concluded that:

> According to a number of companies, the standard of service given by the building industry relates closely to the amount of effort expended by the client in establishing a good brief at the beginning.

Brief execution

It is important that, at subsequent stages of procurement, proper consideration is given to the information assembled at the briefing stage and, furthermore, that it is effectively communicated.

From the design team's perspective, it is usually considered desirable that the individual who will be responsible for the design should be the one to collect the brief from the client. However, effective brief collection and management is a highly skilled business, and not all designers, even the most talented, possess the necessary attributes to do it well. It is important, therefore, that the right person is selected to undertake this task, irrespective of who he or she is.

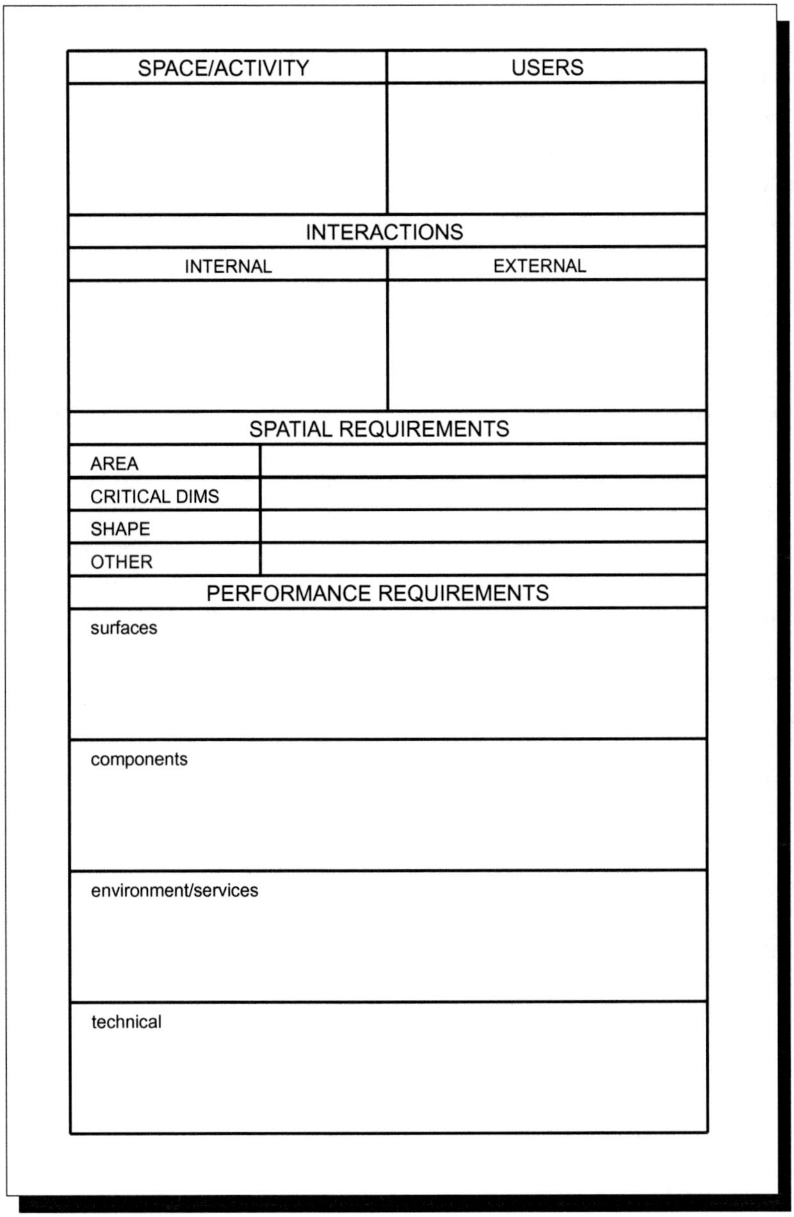

SPACE/ACTIVITY	USERS

INTERACTIONS	
INTERNAL	EXTERNAL

SPATIAL REQUIREMENTS

AREA	
CRITICAL DIMS	
SHAPE	
OTHER	

PERFORMANCE REQUIREMENTS

surfaces

components

environment/services

technical

Figure 4.8 Typical design criteria sheet.

Many large clients now realise the importance of properly developing a brief within the context of their organisation's entire business and financial plan/strategy. In order to ensure that the brief accurately reflects their needs, they increasingly engage the services of a 'professional client' or appoint a project manager. This individual may either be an 'in-house professional', or an external consultant.

In recent years there has been an increasingly flexible approach to building procurement and, as we have seen, a move towards Public Private Partnerships (PPPs). This imposes an increased strain on briefing processes and massively widens their scope, particularly when a design, build, finance and operate arrangement is being entered into. HM Treasury guidance addresses these issues quite specifically in its *Gateway to Success* methodology[7] designed by the Office of Government Commerce. This outlines a procedural approach whereby a project is examined at critical stages in its life cycle before it is allowed to progress to the next stage. At each of these 'Gateways' the project is evaluated against the progress made in addressing the objectives for which it was conceived. The focus is on public sector projects and the emphasis is on the achievement of value for money.

Current procurement developments are very much client driven through a perception that traditional approaches do not give value for money and, whilst they provide opportunities, there is some cause for concern. The main driver seems to be speeding up processes and securing private sector funding, rather than securing an improvement in the performance of the building. However, some play is made of the notion of value engineering in PPP arrangements.

Underlying all these contemporary developments there is also a feeling that, no matter which procurement procedure is adopted, increasing competitive pressures result in cost-cutting being focused on those areas that are perceived as making a less productive contribution. Brief development is one such activity that tends to be marginalised. Many building professionals express some concern that this scenario will leave as big a legacy for the twenty-first century as the mistakes made in the 1960s.

In a similar fashion, as we discussed in Chapter 1, there is still some passage of time needed before the success in complete project terms of PFI and other partnering arrangements can be determined.

Design and construction

Outline proposals and sketch schemes

The next stage in the design process is the production of a sketch scheme. This will, in all probability, overlap with the development of the brief, and the process is very much an iterative one.[8]

All too often a fracture occurs in the logical development of the design, when the concept used for the building is a product of the designer's objectives, rather than those of the client.[9] If proper regard is not given to an accurate interpretation of its findings, the briefing exercise will have been of little value. This strengthens the case for a person other than the actual designer to carry out the user requirement study and write the brief, even though this may be at the expense of more direct lines of communication.

Following the formulation of a sketch scheme comes the important stage at which the designer seeks the client's final approval of the proposals. As the client's perception of the scheme often represents a non-professional interpretation of what is proposed, it is essential for what is on offer to be objectively evaluated against what is required. Here the client will benefit substantially from the advice of an in-house professional or project manager. If accurate judgements are not made at this point then the project may be launched on a course that is misdirected.

Consideration of building performance issues are rarely fully evaluated at this stage. Catt, for example, has pointed out in a series of articles on planned maintenance, the importance of undertaking an early appraisal of design from the maintenance aspect.[10]

Research at Heriot-Watt University[11] indicated that the interest of the design team in maintenance varied over the life of a project, with concern increasing towards the latter stages, where it was largely influenced by economic and practical factors. This confirms a general fear of an irregular and inconsistent approach to the issue.

Detailed design and building defects

It is during the development of the detailed design that there is the greatest perceived scope for technical performance problems to originate. There is a general assumption that poor detailed design practices are the source of a large proportion of performance problems during the life of a building. However, it could be argued that in many cases the real cause lies at an earlier stage.

Defects that occur during the lifetime of a building are all too easily attributed to detailed design, and this may be a hindrance in fairly allocating the blame. Failures, deemed to be the result of fair wear and tear by the expert, may be seen in a completely different way by the building owner. The view taken by the latter must depend upon the original performance expectation. There is, thus, not only a duty on the designer to determine what these performance requirements are, but also a need to make clear to the client what a reasonable level of performance for a given material or component is. Neglecting to do this may lead to many deficiencies that should be classed as defects to be accepted as normal maintenance work, or their rectification as an improvement. It may also lead to the development amongst building owners of a perception about the nature of their buildings that is decidedly unhelpful to a maintenance team.

Normal maintenance, in its strictest sense, ought to consist solely of those actions needed to maintain an element or component in the condition required to perform the function for which it is designed. However, there is evidence to suggest that around 20% of so-called maintenance expenditure can be identified as being necessitated by defects. In general, these tend to manifest themselves early in the life of the building and, whilst the majority are associated with structural matters, there are significant incidences of poor performance of finishes, difficult access problems to services and other detailing problems.

Action to remedy many of these – for example creating additional access to service ducts – may also be classified as improvements, whereas in reality a substantial proportion is work that is necessary to correct design defects.

Detailed studies of samples of buildings and their analyses reveal that the largest single type of fault resulted from making wrong choices about materials or components for a particular situation in the building. The poor decision-making leading to this may have occurred either at the detailed design stage or in the user analysis.

Maintenance determinants can be normal or abnormal when classified on past experiences and current expectations, and each of the following is influential:

❏   the adequacy of the design and the suitability of the materials specified
❏   the standard of workmanship in the initial construction and subsequent maintenance operations
❏   the extent to which the designer has allowed for present and anticipated future needs.

Detailed design decision-making

A major issue, during detailed design, is the selection of materials and components, and choice is becoming increasingly difficult. There are immense pressures tending to change the nature of building from a traditional craft process towards an assembly of factory-made components. This has several consequences. In the first instance, the designer is often faced with having to choose between an extremely large range of products, all of which on face value may satisfy performance requirements. Additionally, designers are subjected to a great deal of lobbying to use one product rather than another, and are bombarded with a plethora of publicity material.

This can have two extreme consequences:

❏   It may sway the designer into using a new component that has not been sufficiently proved in practice, with a consequential failure to meet requirements.
❏   On the other hand, there may be the opposite reaction, resulting in the designer using a component or material with which he is familiar, whether or not it is the right choice in the circumstances.

A number of research projects have concluded that a major shortcoming in detailed design appears to be a failure to make use of the authoritative guidance that was available. This might be a simple failure to use codes and standards properly, or attributable to the sheer volume of advice. In many cases, they suggested, there was what amounted to a perverse avoidance of the standard solution, on the basis that the designer could develop detailed design from first principles.

There is also evidence suggesting that much detailing may be excessively complex, accompanied by over-ambitious specifications, and that the designers' argument that this is in the interests of improved quality is not proven.[12]

The changing nature of building construction, to make greater use of component technology, places much greater emphasis on the need to co-ordinate design from both the technical and the organisational viewpoint. From the technical viewpoint, this manifests itself in a steadily rising incidence of joint failures, which is undoubtedly due to a poor understanding amongst designers of the performance limitations of joints and sealants. This ignorance is mirrored in substandard execution on site and inadequate supervision. All too often it is clear that the designer asks too much of a detail and, furthermore, does not understand the nature and limitations of site operations.

From the organisational perspective, the move towards 'component-based' construction exacerbates a problem that has existed for some time, in that the design and construction of a building is not about the work of only one designer, but that of a team of people working in different organisations. Current trends suggest that this fragmentation is increasing.

The responsibility of the design/construction team

Specialist designers are invariably employed on all but the smallest projects and the task of co-ordinating their efforts, whilst by no means easy, is all too often given insufficient attention. The reasons for this are diverse and cannot be fairly attributed to any one body. There have been a number of views expressed over the years, the most common of which point the finger at the alleged shortcomings of the architectural profession in terms of its management competence. However, the following observations, distilled from a range of commentaries, can be made.

(1) Improving design competence can make the biggest contribution.
(2) Execution on site is the next most important item, and faulty materials/products are a comparatively minor problem.
(3) Lack of design competence is rooted fundamentally in an education system that is divisive and run largely by theoreticians rather than practitioners.
(4) At the level of design procedures, the greatest need is for better co-ordination (design management), and in particular improved checking systems.
(5) Site supervision and the checking of materials and workmanship must be improved.
(6) Possibly the greatest problem faced by the designer is design proliferation.
(7) There is too much unnecessary innovation and not enough reliance on standard, tested solutions of known reliability.
(8) In too many cases there is an absence of a truly effective 'project coalition'.

The client's attitude and understanding must also be questioned as many architects argue that there is rarely time to execute the preliminary stages, or the detailed design work, in the most effective way and that this situation stems from a client's ignorance of the information needs and procedures that must be undergone. Additionally, the financing of buildings, in both the public and private sectors, is subjected to pressures that do not encourage proper consideration of the building's long-term performance.

The building contractor must also shoulder some share of the blame and be prepared to make greater efforts to understand new technology and allocate proper resources to site supervision. Contemporary developments in procurement procedures, whilst often at the client's behest, are not improving this situation, and the widening of responsibility does not help accountability.

There has been a succession of pleas calling for greater involvement of the maintenance manager at the design stage, and also for a more carefully considered view of the whole building procurement process, which recognises building maintenance as an important, integral part of it. A particular criticism in this respect has been of the procedures adopted for handing over, commissioning and running in of the building, and a failure to provide good-quality feedback information. All of these aspects have been the subject of frequent recommendations over the years, but infrequent action by the industry, representing a major failure in terms of the service it provides to its clients.

The Construction (Design and Management) Regulations 1994

Health and safety on construction sites has been an important issue for some time, with the major responsibility lying with the contractor. The Construction (Design and Management) Regulations 1994 (CDM)[13] have had an important influence on both the execution and management of maintenance. This is because they introduced a major transfer of responsibility for important aspects of safety from the contractor to the designer and building owner.

Determining when the regulations apply is rather complicated, but in simple terms they are applicable to all but the smallest jobs. These are defined as jobs that:

❑   will not be longer than 30 days
❑   will not involve more than 500 man-hours of work
❑   do not employ more than four people at any one time.

In-house operations do not come within the scope of CDM. Work to a person's own house is also exempt, provided it is not part of a trade or business. However, demolition works are always subject to CDM, unless for a householder or in house. CDM is always applicable to design work, no matter how small the job.

There are also rules to determine what constitutes a job or project. Under the regulations, the project ends with hand-over and occupation. Operations subsequent to this constitute a new project.

The European Directive on Temporary or Mobile Work Sites[14] defines the term 'project supervisor', and under the UK CDM regulations this role is divided into two, with responsibility resting with 'the planning supervisor' and the 'principal contractor'.

There was an early recognition that no one professional was necessarily equipped to fulfil the role of planning supervisor and this has given rise to the creation of a specialist consultant. The planning supervisor is appointed by the client to:

❏ notify the Health and Safety Executive of the existence of the project
❏ ensure that designers fulfil their responsibilities under CDM
❏ ensure that designers co-operate on site safety matters
❏ ensure that a health and safety plan is produced
❏ ensure that a health and safety file is produced and amended as necessary, and handed to the client on completion
❏ give advice on appointments.

The designer's responsibilities are to:

❏ inform the client of his (the client's) responsibilities
❏ avoid or reduce foreseeable risks on site
❏ ensure that the design has adequate information on health and safety
❏ co-operate fully with the planning supervisor.

The approved code of practice[15] gives a great deal of advice with respect to the interpretation of these duties. Furthermore, there is an acceptance that health and safety has to be measured in the context of overall building design and construction requirements and, in particular, economic realities. A cornerstone of the requirements is the need to carry out risk assessments. For a designer this means identifying the hazards in a proposed design and designing them out wherever possible. Where it is not feasible to eliminate the hazard the designer must consider suitable control measures to reduce the risk during both the construction and occupation phases of the building life cycle.

Under the designer's duty to produce designs that can be safely built, the most important consideration will be access, not only during construction but also for cleaning, servicing, maintenance and repair. The client must automatically receive a health and safety file at hand-over, warning of any potential risks, and the designer is responsible for providing the planning supervisor with the information to go in it.

Whenever the CDM regulations apply there must be a client, and this is defined in the regulations as 'any person for whom a project is carried out'. There are no requirements in respect of client competence, nor for adequate resourcing (unlike for the appointment of designers, safety supervisors, principal contractors and other contractors).

The clients responsibilities are to:

❏ appoint a planning supervisor and a principal contractor
❏ ensure that the contractor is a contractor under the definition of the regulations
❏ ensure that the planning supervisor is competent under the provisions of the regulations
❏ ensure that the appointees are adequately resourced
❏ ensure that the health and safety plan exists before the work commences
❏ ensure that the planning supervisor has information about the condition of the premises
❏ ensure that the health and safety file is available
❏ pass on the health and safety file to anyone acquiring an interest in the property.

Clients can delegate all or any of these duties to an agent, and the major role of the planning supervisor in advising the client must be borne in mind. Of prime importance is the requirement to provide the planning supervisor with information relevant to his duties, and this imposes a major obligation to maintain accurate and up-to-date building records which will include:

❑   previous safety files
❑   condition survey information
❑   maintenance arrangements
❑   details of plant and equipment
❑   occupational information, both past and present
❑   covenants and tenancy information.

At the time of writing the HSE is proposing to consolidate the current legislation in the Construction (Design and Management) Regulations 1994 and the Construction (Health, Safety and Welfare) regulations 1996 with a view to implementing revised CDM regulations in Spring 2007.

Life cycle costing

Building owners, as a rule, place undue emphasis on the capital costs of a project at the expense of future running costs. If a proper balance is to be struck, this issue needs to be addressed whilst the brief and the detailed design are being developed. Various techniques have been developed to assist in the analysis of a building's total costs over its life span.

The life cycle cost (LCC) of an asset is defined as the present value of the total cost of that asset over its operating life, including initial capital cost, occupation costs, operating costs, and the cost or benefit deriving from disposal of the asset at the end of its life.

The broad objectives of a life cycle costing exercise may be:

❑   to enable investment decisions to be made more effectively, taking into account all costs that may arise from it
❑   to consider the impact of all costs, rather than just capital costs
❑   to provide information that can contribute to the more effective management of the completed building
❑   in the context of building procurement, to assist in the evaluation of alternative solutions to specific design problems.

For example, the technique may be applied to an element, such as a flat roof, where there is a need to make a decision between a high initial cost, low-maintenance, long-life solution and a low capital cost, high-maintenance, short-life one. This may be carried out using discounting techniques to compare present and future costs for each alternative on a common basis.

The important point to note about the use of such a basic example as this is that its evaluation is far from simple. Indeed, such evaluations are complex, and a proper analysis requires consideration not only of direct costs and/or benefits, but also indirect ones. Some of these may readily be quantifiable, such as disruption costs owing to failure and/or repair execution, whilst others will require more subtle forms of evaluation. An example of the latter may be an aesthetic consideration in a sensitive part of a building, such as the reception area of an important office facility or hotel.

Sound judgement is needed when calculating life cycle costs, together with the exercise of technical, managerial and financial skills.

Implicit in the term 'life cycle' is the notion that, during its life, a building progresses through a number of phases. The nature of the cycle will be dynamic, as opportunities for change, renewal and adaptation present themselves, and this must be recognised when calculations are being made.

The sequence of the life cycle phases is described, in engineering terms, in BS 3811, and for the life of a building are:

❏ brief collection and development
❏ design
❏ construction
❏ commissioning
❏ maintenance
❏ modification
❏ replacement.

An essential element in using LCC is defining the life cycle period, or 'building life', and within this, component and/or element lives.

Life cycles, obsolescence and utility value

Buildings are, in general terms, very durable, and if properly maintained may last for centuries. Even if maintenance is indifferent, a physical life of 50–60 years is quite realistic. The term 'physical life' here means the life of the identity of the building, and not necessarily the life of every physical part of it. The attitude of the client to the required physical life of the building may be established in preparing the brief, although it might not be the subject of a specific statement.

Many components will need replacement during the life cycle, some several times. The length of time for which it will be worthwhile to continue to repair and renew parts of a building will depend on how well the building continues to meet the needs of the function for which it was built. In economic terms, past costs are of no importance, and only the relationship between future costs and future value is significant. A building is worth repairing if the future utility to be derived from it exceeds that which could be obtained by demolishing it and erecting a new one.

In an extreme case, it can be worthwhile demolishing a new building when, for example, the purpose for which it was built has ceased to exist. It is worth noting that this situation might arise as a result of a bad initial investment decision by the client, an inadequately managed inception process, or perhaps a poor response by the professional advisors. At this point, of course, the possibility of adaptation may be considered, and in a sense this could be said to give the building a new life, in that its identity has been changed.

Any consideration of building life must take account of the notion of obsolescence, which relates to economic considerations, directly or indirectly. The RICS has identified the following six forms of obsolescence:[16]

- ❏ economical
- ❏ physical
- ❏ functional
- ❏ technological
- ❏ social
- ❏ legal

Economical and functional obsolescence are probably the most common of these. However, all these forms are interlinked, and the simpler categorisation of Ashworth into physical, functional and economic lives may be more realistic.[17]

As obsolescence is concerned with the potential utility that may be derived from a building, some thought must be given to the rather contentious issue of value. In the commercial sector this may be something that is clearly objective, and value can be stated in money terms, with the value of the building being determined by its worth as a factor of production. In the service sector the notion of value is much more difficult to isolate, and it is often here that building fabric performance is of central importance. Value, therefore, is determined by not one, but many attributes.

The assessment of a building life is a difficult exercise, requiring good judgement, knowledge of context and previous records. For business accounting purposes, the building life is the period over which the relevant organisation holds an interest in the building. At the end of this period the building may have a residual value and pass into the interest of another organisation.

The application of LCC must always be carried out in a proper context, as considerations of a life cycle may differ somewhat from notions of the life of a building, or a component as normally perceived. Component life may conceptually be rather easier to define, if not to evaluate, as it will generally be shorter and more easily foreseeable than that of the building itself. The circumstances leading to the replacement of a component may include deterioration, failure or obsolescence, and the last mentioned should be considered separately from the others. The issue of component life is a very important one for its own sake, and not simply as part of a life cycle costing exercise.

Life cycle costs

Life cycle costs can be divided broadly into three categories.

(1)   Capital costs which, in theory, may be relatively easy to predict.
(2)   Costs-in-use which are more difficult to predict, and include several components.
(3)   The costs involved in final clearance of the site or disposal of the asset.

Against these costs can be offset the residual value of the asset: that is, financial inflows accruing from disposal. A simple view therefore is:

LIFE CYCLE COST = CAPITAL COST + COST-IN-USE − NETT RESIDUAL VALUE

Maintenance costs will only be part of the cost-in-use, and all recurring costs must be fully considered. Generally the three R's of running, repairs and replacement will cover costs-in-use, and these may all be subject to consideration during the design process.

Exhaustive lists are difficult, but the following will typically need to be counted:

❏ maintenance, including redecoration
❏ energy consumption
❏ cleaning
❏ rates
❏ insurance
❏ estate management or management overheads
❏ finance costs.

There is also some debate as to whether the costs of alterations and adaptations should be considered, and whilst this is a difficult issue in definition terms, there is no doubt they must be part of the life cycle cost equation, irrespective of whether or not they were predicted. For practical and operational purposes, and the factors discussed above, it is not possible, or indeed realistic, to completely ignore the question of defects, repair/replace decisions and improvements that will be necessary to maintain the utility of the building.

Expenditure made to materially change the function of the building – that is, to change its identity – will, however, be relegated in significance.

Financial appraisal techniques

The analysis of costs and benefits

For accounting purposes, it is convenient to divide costs into capital and revenue. Expenditure in each of these categories is often determined by different decision-making processes, and this is a major obstacle to the efficient implementation of LCC. The main life cycle costs are summarised in figure 4.9.

One of the purposes of carrying out an LCC exercise is to assist a decision-making process during design. This may be at the global level, when strategic design decisions are being made, as well as at the detailed design stage, when it is necessary to decide between the relative merits of different specifications, and where relative life cycle costs will help inform the choice.

There are many techniques of financial appraisal that can be used, the simplest of which is the payback period approach. This is the time taken for the returns from an investment to equal the initial outlay, and is sometimes used to evaluate the financial benefits of a variety of energy conserving measures. For example, if roof insulation

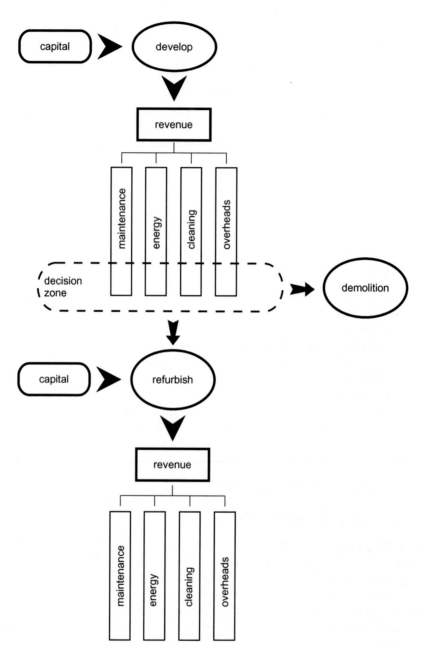

Figure 4.9 Life cycles and life cycle costing.

costs £200, and it is estimated that the measure will reduce fuel bills by £40 per annum, the payback period is five years. The major weakness of this method is that it does not take into account the time value of money.

LCC techniques are concerned with the evaluation of a stream of costs over time and with present value. This implies that to evaluate costs properly there must be a common basis of comparison, and future costs need to be converted to present-day equivalents by discounting them. Appendix 2 explains the principles involved in the discounting of future costs, allowing them to be compared with expenditure now, on a common basis.

The illustrative example given evaluates the relative merits of two alternatives, A and B. The former involves a capital outlay of £100 000, with a predicted annual maintenance of £2000. The alternative has a capital cost of £120 000, with annual maintenance expected to be £500 per annum. If the time value of money is ignored, then alternative A involves a total outlay of £220 000 and B £150 000.

If the calculation is repeated, taking into account that a £1 in the future is worth less than a £1 today, then the picture is not so simple.

In the example given in Appendix 2, a discount rate of 9% over a period of 60 years suggests that a comparison on a present-day value basis makes A the cheaper option.

The calculations can be performed in three main ways, but all are based on the same principle.

(1) The simple example referred to above discounts all future cash flows to give a NETT PRESENT VALUE.
(2) Alternatively, the cash flows over the life of a building or component can be converted using discount tables to an ANNUAL EQUIVALENT COST.
(3) The third method that can be used is to determine the interest rate, which when applied to all future cash flows gives a nett present value of zero. This is then called the INTERNAL RATE OF RETURN.

For purely comparative studies, either of the first two methods may be used, but the use of an annual equivalent may be of additional benefit for financial management purposes.

The internal rate of return may be a useful measure in commercial decision-making, but is of less relevance for present purposes.

In terms of the design decision-making process, there are several characteristics that need to be examined with respect to the discounting techniques.

(1) The techniques assume that the designer is choosing from mutually exclusive options. In practice he may have a larger range of options available that are not mutually exclusive. Design decisions are rarely a simple either/or choice.
(2) Only negative cash flows are being considered, and the positive benefits of each, whether or not they can be evaluated in cash terms, are not included. Any comparison purely on these grounds is assuming that each option gives the same utility. Public sector organisations, for example, have many demands to satisfy, which will include social and legislative factors. In such cases the use of cost

benefit analysis may be desirable. In the private sector, too, not all benefits and costs can be evaluated in pure cash terms, although it may be more realistic to attempt to measure utility in a theoretical manner.

(3) It should also be pointed out that a fully rigorous life cycle analysis should count all costs, including those that stem indirectly from maintenance operations, such as the cost of disruption to the everyday performance of the building.

(4) The use of the techniques assumes the cost of maintenance expenditure is determined on the basis of predicted needs and, given the inevitable pressures on maintenance funding, these might not be fully met. The techniques described might show that a low-cost, high-maintenance solution is preferable under some circumstances. However, the question has to be asked whether in reality funding for the maintenance work will be forthcoming when resources are scarce. The consequent deterioration of a building or element, if not properly maintained, may severely impair the performance of the building to such an extent as to render the original analysis fatuous.

Inflation

So far no account has been taken of inflation and there are several reasons for ignoring this, including the difficulty of predicting it over a building's life. For comparative studies, it is assumed that inflation will affect each option in the same way and can therefore be ignored. This is rather questionable, as the error attached to proceeding on this assumption may well be greater than has hitherto been assumed. For example, under some market conditions labour costs may rise at a greater rate than material costs, and as maintenance tends to be more labour intensive than new build, substantial errors may accrue.

In the final analysis, however, it is clear that any attempt to allow for inflation will greatly complicate the process, and given the range of uncertainty attached to all the input data, it is probably wise to avoid a too sophisticated approach.

Uncertainty and data quality

All the techniques require the forecasting of future cash flows and are, hence, subject to substantial uncertainty. It has been argued that, for comparative purposes, this may only have a limited influence. It is also true that the greatest uncertainty attaches to long-term cash flows, and the examples in Appendix 2 show that these will be subjected to very small discount factors, so that the margin of error is correspondingly smaller.

What is certain, however, is that data on the life of a building and its components is an essential pre-requisite. Such data is difficult and expensive to obtain and to update, particularly if it is to have statistical significance.

A summary of appropriate statistical techniques that may be useful in assessing maintenance data is given in Appendix 1.

Economic life

All discounting techniques require an assumption to be made about the lifespan of the building under consideration. Traditionally, a 60-year life has been taken for analyses such as those discussed above. The points made earlier with respect to notions of economic life and obsolescence must, however, be borne in mind. One retail chain, for example, takes an economic life to be 11 years, because it is assumed that at the end of this period the market will have either grown or disappeared, thus requiring either the disposal or replacement of the building. This is becoming increasingly true of many modern industrial buildings.

The notion of the disposable building has become more readily accepted than hitherto and investment decisions have been adjusted accordingly. In such a scenario it is reasonable to expect this requirement to be an important aspect of user requirements, which should be clearly identified during the collection of the brief, and responded to accordingly. Current sustainability debates, however, challenge these purely economic-based decisions and increasing environmental pressures will force a more holistic analysis of building life.

Interest rates

LCC techniques also require the prediction of interest rates, and this is an area of undoubted concern. The higher the interest rate, the lower the present value of future costs. In the comparison example given earlier, a 9% discount rate rendered option A, with a lower capital outlay, the cheaper. At a 6% discount rate, option B becomes the cheaper option in present value terms. It can be seen, therefore, that high interest rates favour lower capital outlay, as future maintenance costs appear less and less significant (figure 4.10).

This is a serious weakness because, as has been seen, the analysis tends to ignore many unquantifiable factors which may seriously affect the building's performance. Not least of these may be the energy implications in material usage of high-maintenance buildings, not normally taken into account in discounting techniques, because of a lack of usable data.

The effect is even more pronounced when inflation is taken into account, as maintenance and running costs may be set against taxable profits, which may to some extent mask inefficiencies in the asset's performance. More sophisticated analysis of the effect of interest rates can be carried out using sensitivity analysis.

If a 15-year life and a 10% interest rate are taken as the base case, and a vertical and horizontal line drawn through this point, some measure of the sensitivity of the analysis to changes in the variables can be obtained. For a given interest rate it can be seen that, as would be expected, a reduction of asset life causes a steep increase in annual costs. Similarly, an increase in life causes a reduction in annual cost, but at a reducing rate.

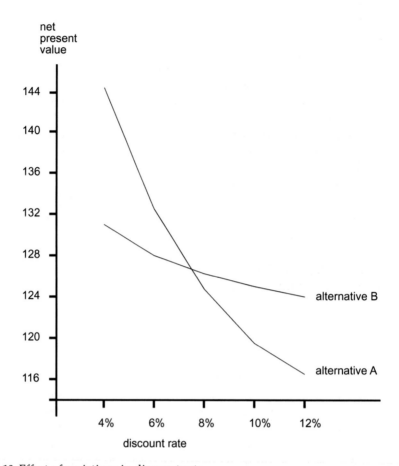

Figure 4.10 Effect of variations in discount rate.

For a given life it is also easy to examine how annual costs vary with interest rates. Note that, at a ten-year life, an increase in interest rates from 10% to 20% increases annual costs by approximately £850. However, at a six-year life, the same rise in interest rates increases annual costs by £750.

Application of techniques

The basic concept is that design decisions should take into account long-term costs, as well as initial costs, and that the overall design solution should strike the right balance between these. The technique can be used at all stages in the process.

At inception stage it may be used as a tool to evaluate a range of strategies for satisfying the demand for building space: for example, in deciding between refurbishment of an existing facility or construction of a new one. At the briefing stage the technique can be used as part of the two-way dialogue between designer and client, in order to assist the client in this difficult decision-making process. Indeed, determination of the relationship between initial expenditure and running costs, at least in principle, is an essential component of the brief. If the briefing stage has resulted in providing a proper framework for design, then LCC is an important tool throughout the whole of the detailed design phase.

Furthermore, during building occupation LCC can also be used to aid decision-making concerning the framing of maintenance and renewal policies, and as part of the overall system of financial management and control.

However, these techniques should never be considered as a panacea for the resolution of all difficult decisions. They are simply aids which, together with others, can be used to supplement skill and judgement.

Value engineering

Value engineering is a technique that evolved in manufacturing engineering in the USA and there are a range of definitions:[18,19]

❑ The creative organised approach whose objective is to optimise cost and/or performance of a product or service.
❑ A disciplined procedure directed towards the achievement of essential function for minimum cost without detriment to quality, reliability, performance or delivery.
❑ The systematic review and control of costs associated with acquiring and owning a facility or system.
❑ The systematic application of techniques which identify the function of a product or service, establish a monetary value for that function, and provide the necessary function reliably at the lowest overall cost.

Although these definitions imply attention to cost, they identify that value engineering differs from conventional cost control in that rather than focusing on simple cost it concentrates on value.

The construction process

There has been an ongoing commentary, over a number of years, concerning the standards of supervision, and hence quality levels, in the buildings produced by the UK contracting industry. Therefore, the contracting side of the industry should be subjected to the same critical scrutiny as the design side.

The move towards buildings becoming an assembly of components not only has repercussions for designers, but also places a series of different demands, both technical and organisational, on contract management. In the technical sense, the issue of joint performance is an obvious one, and the requirements this imposes on assembly, accuracy and tolerances are well understood in principal.

At the more specific level of detailing, the ability of the designer, through his knowledge and experience, to provide technical information that is both functionally correct and practical is of prime significance. What is most important, however, is the need to recognise that proper solutions to these problems do not rest with either designer or contractor alone, but require each party to have a proper understanding of each other's problems and objectives.

The organisational consequences of the changing nature of contracting impose a much broader set of problems, and these stem from two factors.

(1) The changing technology of buildings increases the number of parties to the contract. For example, the use of particular components may introduce an additional specialist at the design stage and an additional specialist contractor during construction.
(2) The motivating force behind current practices in contractual arrangements have not been driven by building performance objectives, but rather by the desire to reduce procurement time.

Whilst there may be some hidden benefits in more flexible contractual relationships in terms of performance, it is undoubtedly true that the role of the main contractor has diminished to the point where his role, more often than not, is essentially that of co-ordinator of a series of work packages. Management contracting formalises this situation, but even on contracts let by a traditional competitive tender, the main contractor employs few, if any, of his own operatives on site. All the work is likely to be subcontracted, to either domestic or nominated subcontractors.

Information management

In response to the Latham report and a report by the Building Economic Development Committee in 1987, the Co-ordinating Committee for Project Information (CCPI) was established. Their mission was to respond to findings indicating that the largest single cause of quality problems was inadequate project information.

Despite the work of CCPI, the quality of information management still remains a grave cause for concern and the essential link between design and construction needs to be recognised, as readily available, reliable and easy to understand project information is critical to effective pricing, planning and control.

The following deficiencies in project information were identified by CCPI:

❏   missing
❏   late
❏   incorrect
❏   insufficient detail
❏   impractical
❏   inappropriate
❏   unclear
❏   not firm
❏   poorly arranged

❏   unco-ordinated
❏   conflicting.

Studies of quality levels on building sites have concluded that the predominant causes of quality problems on site were poor workmanship and unclear or missing project information, but that a proactive attitude by the contractor towards information flow leads to higher levels of quality being achieved.

The link to quality is two-fold. In the first instance, construction information produced hurriedly under pressure is more likely to be of a poor standard. Secondly, there is massive disruption to site processes, with inevitable consequences for quality. The site manager is frequently being deflected from his role as site supervisor by having to spend a disproportionate amount of time trying to rectify information deficiencies.

The problem, however, is not only one of information deficiency, but one of poor co-ordination between sets of information. This will also lead to an enormous amount of time-wasting on site in carrying out detective work to elicit the answer to a technical query, and/or to resolve conflicting information. Information conflicts themselves are a manifestation of poor co-ordination between members of the design team.

CCPI concluded that amongst a number of construction industry problems strongly influenced by information problems were:

❏   the incidence of technical defects in finished buildings
❏   the frequent poor quality of finished work, related to high maintenance costs.

In 1987 CCPI published a comprehensive guide to the co-ordination of project information.[20]

CCPI has now been reconstituted as the Construction Project Management Information Committee (CPIC) which in 2003 launched a new code, *Production Information: A Code of Procedure for the Construction Industry.*[21] A significant addition to previous documentation, it recognises the huge changes that have taken place in information technology, especially CAD systems and computerised specification systems. It acknowledges the potential for improved co-ordination and information-sharing but also highlights the importance of good strategic management. It includes robust guidance on effective file-structuring without which the potential advantages will be negated.

As well as addressing the primary issue of quality of production information management, it is of importance relevant to the broader issue of maintenance information management. It imposes structure on information, allowing more effective annotation and cross-referencing, which is of particular benefit in the production of as-built drawings.

A limited survey at De Montfort University, Leicester in 1992, suggested that only 55% of respondents were aware of the recommendations of the original CCPI, and only just over a third had adopted common arrangement principles.[22] The reasons given for this were various, but generally revolved around a reluctance to change existing practices. At the time of writing there is no comprehensive research as to the extent of current 'take-up' although anecdotally it is accepted that the majority of design offices now use CAD-based drawing systems.

Hand-over and commissioning

The roles and responsibilities of the various parties to the project continue into the occupation of the building, and it is pertinent to consider the information needs of the person taking delivery of the building. *The Architects Job Book*, in stage K, includes an item requiring the architect to remind all relevant personnel to provide the necessary information for the preparation of a maintenance manual. It is a matter of some conjecture as to how often such a manual is actually produced.

Rooley comments that:

> Commissioning and the attainment of the Practical Completion Certificate are therefore viewed from opposing standpoints by the designer and facilities manager. One has the expectancy of relief and escape from the project, and the other anticipation and excitement of a new beginning.
>
> Hand-over documents should include the design intent, an accurate set of as-fitted drawings, a set of commissioning figures and modifications carried out during commissioning and a set of maintenance instructions which are specifically relevant to the building and to the operation of each component in context.[23]

The CIOB refer to project information manuals.[24] Such a manual consists generally of two elements. One part will be concerned with provision of information to the building owner to enable him to use and manage the building properly, and the second part will consist of instructions for its management, with the broader facilities management scenario in mind.

Contents of the building manual

(1)   Information

The most essential component is a set of as-built drawings. The *RIBA Job Book* advises that these should be specially prepared. There are two major reasons for this. In the first instance, even the latest most up-to-date set of production drawings are unlikely to be as-built, as all instructions and variations will not necessarily be included in a revised drawing, and because of possible discrepancies between drawings and construction. Secondly, as-built drawings need to be produced for the very specific purpose of assisting with the proper running of the building, for both operational and technical purposes. Any attempt to 'fudge' the issue by the presentation of a pack of rather dubious information to the building owner will be counterproductive.

Given the huge growth in CAD-based production systems, it would seem that there is no real excuse for failing to have on hand as-built drawings that are up to date and, given the power of the technology available (e.g. layering systems), in a format that is of real use to the building manager. Presentation in this form also provides the means by which this information can be continuously kept up to date.

Possession by the building manager of a relevant and accurate set of as-built drawings is essential at a variety of different levels.

❏ At the strategic level all good maintenance management systems require accurate and up-to-date records of the estate. The as-built drawings, provided at hand-over, are only the first step in this process, as a continuous up-dating process is necessary to incorporate additions and alterations.
❏ These drawings need to be appropriately annotated for the specific purpose of managing the building. For example, information for fire safety and the needs of maintenance management require a particular format.
❏ At the operational level there will be the need for detailed information with respect to specific parts of the building system. An important example is the need to provide drawings identifying service routes, and the means of accessing them.

There is no standard pattern for the production of these drawings. The best advice is to stress that their preparation should be considered as a logical development of the project information system. The drawings should be properly co-ordinated with other information sources (see CPIC), because the drawings will need to be supplemented by other information, which may be in the form of schedules or tables.

The manual should contain all the necessary information about the construction of the building, and therefore comprise not only as-built drawings and schedules, but all the necessary information to ensure that proper management of the building, in the widest sense, can be effected. It is impossible to present an exhaustive list at this point, but each of the following aspects is worthy of consideration.

❏ A properly produced set of as-built drawings supplemented by relevant schedules, notes, and specifications of materials and components used in the construction of the building.
❏ A reference section, where all relevant sources of information are given. This may include manufacturers' and suppliers' data, such as spare part numbers for plant and equipment, and guarantee or warranty information. British Standard references may also be relevant.
❏ A detailed breakdown of the parties involved in the design, construction and commissioning of the building. For example, the addresses of consultants, subcontractors, etc. should be included, along with points of contact.
❏ Historical data with respect to defects and their making good in the form of a log book may provide important intelligence for the building manager.

If one adds to this collection of items other information required for managing the building, it very quickly becomes apparent that the information content of the manual is considerable, and in itself represents an information management problem of some complexity.

(2)   Instructions for the use of the building

The need for occupiers to be aware of how to use a building properly has been high-lighted in studies of energy usage. At the level of the simple dwelling, it has been found that sophisticated energy-saving designs are of no avail unless behavioural aspects are taken into account. This is clearly a significant factor when complex active service systems, incorporating comprehensive control devices, are used. Occupants must be adequately instructed as to their use.

There is a message here for designers, both in the assumptions they make when designing a building, and in the need to explain to clients how the building has been designed for use. For example, many modern dwellings incorporate conservatories as a sunspace. Whilst the theory behind them is perfectly sound, their benefits are negated when occupiers use them as an additional room all the year round, which entails heating them during the winter.

The building owner also needs to know the maintenance requirements of the build-ing in terms of what to maintain, when to maintain and how to maintain. However, it can be difficult to determine the level of detail that is required. This would be much easier to resolve if matters such as the organisation's attitude to maintenance policy and the nature of the building's usage have been accurately identified at the briefing stage.

For example, the building may become part of an established estate, with a clearly worked-out policy in terms of planned maintenance, and the building manual would be written with this in mind. In such a case it might be entirely superfluous to produce a comprehensive maintenance guide in terms of planned maintenance recommenda-tions. The manual can simply refer to an existing estate document, in which case only those maintenance features peculiar to the building under consideration need to be identified and accompanied by specific instructions.

On the other hand, it may be felt that the building requires its operations manual to give full and comprehensive recommendations for all its maintenance operations. However, to indulge in such an operation, as part of the manual, is probably a mistake as it will lead to the manual's becoming far too cumbersome. There are exceptions to this, the most common of which relates to the mechanical services in the building. As well as making a heavy demand in terms of information content for the manual, it is this subject that will require the most attention in the instructions section of the manual, which has to be read in conjunction with the information section. Very careful structuring is essential to permit easy and accurate cross-referencing.

Preparation of the manual

Two problems immediately present themselves. Firstly, how this information should be collected, and secondly, how it should be structured to facilitate its use by a diverse range of users.

Part of the answer to each of these questions depends on the way the project has been managed in its earlier stages. There are undoubted benefits to be derived if the collection of information for a building manual is seen as an ongoing process throughout the project. Indeed, one could argue that the first set of data for inclusion, produced at the outset, is a proper set of user requirements.

The importance of the manual should be stressed, not only for its essential utility but also for its tendency to act as a catalyst. At the brief development stage, the knowledge that such a manual is to be produced provides a great stimulus for proper consideration of how the building is to be managed, and therefore by implication the needs of the building and its users. At hand-over, the assembly of a package of information to accompany the building also encourages a critical analysis of what has been achieved. This can help feedback, but more importantly perhaps, if the preparation of the manual has been an integral part of the whole process, it should have led to a more considered approach to both design and construction, from the point of view of the eventual user.

Structure of the manual

There is no set formula for the structure of the manual. The two major components will, however, certainly require further subdivision, and there is a strong case for considering the information portion in two parts. This should consist of essential working documentation, such as the as-built drawings and schedules, which may then be further subdivided into sections representing parts of the building or site.

More detailed reference material, such as manufacturers' data, may best be included in a specific reference section, which can be accessed from both of the two major sections. The case for separating out the reference section is a strong one, as this may be organised in the manner most appropriate for referencing from the other sections, rather than subdivided according to parts of the building.

The instructions may be subdivided into either parts of the building or site, or into elements. A case may be made for either approach, and the decision in this respect will depend on the particular characteristics of the project. For example, in a housing project an elemental approach is most appropriate, whereas on a complex multi-functional development subdivision may best be into sections of the project, or perhaps building types.

To summarise, then, a typical manual would consist of the following:

(1) instructions for use of manual
(2) information section
  ❏ general information
  ❏ as-built drawings and schedules
  ❏ basic materials and components
  ❏ parties to the project
(3) instructions for operation of the building
  ❏ general buildings management
  ❏ maintaining the building fabric
  ❏ maintaining the building services
(4) reference section.

Presentation

The way in which this information is presented is open to choice. The use of loose-leaf binders is convenient and facilitates easy updating, but given the ease and flexibility of modern computer software, an electronically based format has obvious merit, not least in terms of its ability to be:

❑   updated easily
❑   searched
❑   accessed by non-cognate users
❑   linked to a help desk facility.

Linkage to databases, to form an integrated system, creates further opportunities for the development of an automated building model, which may encompass full asset registers.

The help desk is another development where individual facilities are linked to a central information source, and is designed for use by non-professional as well as professional staff. Future developments are likely to lead to the use of expert systems to help manage buildings, and these systems may also provide an excellent tool for providing feedback to design teams.

Maintenance and the building manual

A disciplined approach to the preparation of a building manual should generate the following benefits.

(1)   Enable a better dialogue to exist between the designer and the maintainer, and also have the effect of promoting some feedback to the design team. If the maintainer has been involved at the brief development stage, as is advocated, then the preparation of the manual should follow quite naturally.
(2)   Enable a property to be maintained more effectively, both in the organisational sense, and to the proper technical standard.
(3)   Enable and encourage the building owner to plan the effective maintenance of the building, in terms of both planning maintenance programmes, and assisting in the formulation of budgets. It may, in addition, provide an important tool for the maintainer when he seeks maintenance funding.
(4)   By helping to ensure that the building is used properly, it contributes to the reduction of avoidable maintenance tasks.
(5)   As a discipline for the design team, it fosters a more rigorous consideration of the effectiveness of the building in use, and may encourage a critical appraisal of how well intentions were defined, and then met. In this sense it provides some sort of testing ground for the assumptions made at the earliest stage of the project. If the original brief is developed in full expectation of a building manual being produced at hand-over, this will undoubtedly act as an important stimulus to give more complete consideration of performance, including maintenance. Similarly, during detailed design it may serve to encourage a rational approach.

The preparation of the manual does not, of itself, ensure a longer-term involvement of the design team in its long-term performance, but does act as an important focus at a critical phase in the process. The question of this longer-term involvement by the designer has always been a tricky one, and in reality is probably not a feasible option.

It is all the more important, therefore, that the essential continuity is provided by some other person. This may be an in-house professional client, when one is available, or, failing this, the maintainer or maintenance manager.

Maintenance feedback and post-occupancy evaluation

Based on a database of building occupation problems he has collected over a 15-year period, Rooley discovered that 60% of the buildings studied had apparent maintenance problems as a result of:

'fuzzy edge disease' where there is a lack of communication among the various parties in the design, procurement and maintenance process.[25]

This highlights the essential communication problem identified earlier, and the provision of sensible and reliable feedback information is one component that should be considered.

It is self-evident that without feedback from the industry as to how buildings have performed in the past designers are unable to determine how their design decisions will work. This may take a variety of forms but, for it to be effective, there has to be:

- ❏ an incentive to prepare it
- ❏ an expertise to analyse it
- ❏ a system to store it
- ❏ a readiness to use it.

Maintenance activity generates vast quantities of data, and there are efficient means available for its storage and analysis. Although this data may not be feedback-specific, contemporary data management systems permit a range of sophisticated analysis tools so that this should no longer be a primary obstacle.

It is important, also, to recognise that the information required relates not only to detailed performance behaviour but to the validity of building concepts. The most effective way of addressing this is the involvement of building managers in the design process, preferably at briefing stage, and in some large estate-owning organisations this may be possible.

There is, however, a need for more published information, in the right format, and for this information to be taken note of:

Feedback information is a continual problem and further efforts are required to ensure that problems found during maintenance works to have resulted from incorrect detailing, poor specification and choice of materials, defective construction techniques and workmanship, are not repeated, as these prove to be very costly. In this connection it would be necessary to record these defects and also to record the remedial measures which have been taken.[26]

Some steps have been taken through more attention being paid to evaluation of buildings in use through post-occupancy evaluations (POE).

Linked to their Gateway process, the Office of Government Commerce, in its procurement Guides, recommends post-project reviews where an evaluation is carried out against established key performance indicators (KPIs).[27, 28] The all-industry KPIs for the construction industry include broad indicators for quality, but these are performance-related and as such do not provide analytical data at the appropriate level of detail. Following this, it advises the execution of a post-implementation review (PIR) when the building has been in use long enough to determine whether the original objectives have been met, and it is at this level that the generation of feedback data becomes significant.

The design quality indicators toolkit also includes an in-use module. It emphasises that in carrying out this type of exercise the use of a properly trained facilitator is essential and the Construction Industry Council (CIC) maintains a list of accredited people.

There is therefore more good work being carried out in evaluating finished projects, but the question remains as to whether the feedback is being completed in terms of the right lessons being learnt in conceiving new buildings.

References

(1)   LAMSAC (1981) *Terotechnology and the Maintenance of Local Authority Buildings.* LAMSAC, London.
(2)   British Standards Institute (1984) *BS 3811: 1984 Glossary of Maintenance Management Terms in Terotechnology.* HMSO, London.
(3)   O'Reilly, J. N. (1987) *Better Briefing Means Better Buildings.* HMSO, London.
(4)   www.adb.dh.gov.uk
(5)   www.dqi.org.uk
(6)   Markus, T. A. (1977) *Building Performance – Building Performance Research Unit, Strathclyde University.* Applied Science Publishers, Barking.
(7)   Office of Government Commerce (2005) *Gateway to Success.* Office of Government Commerce, London. www.OGC.gov.uk
(8)   Powell, J., Cooper, I. & Lera, S. (1984) *Designing for Building Utilisation.* E. and F. Spon, London.
(9)   Stonehouse, R. (1983) Housing in the Humanist age, *Architects Journal,* 9 February.
(10)   Catt, R. (1986) Bringing buildings into care, *Building Technology and Management,* 10 July.
(11)   Spedding, A. (ed.) (1987) *Building Maintenance and Economics – Transactions on the Research and Development Conference on the Management and Economics of Maintenance of Built Assets – Maintenance considerations during the design process.* E. and F. Spon, London.
(12)   *New Builder* (1993) Complex specifications force up UK building costs, *New Builder,* 16 April.

(13)   Statutory Instruments (1995) *Health and Safety: The Construction (Design and Management) Regulations 1994.* HMSO, London.

(14)   European Commission (1992) *Directive on Temporary or Mobile Work Sites – 92/57/EEC.*

(15)   Health and Safety Executive (2001) *Managing Construction for Health and Safety: Construction (Design and Management) Regulations 1994 – Approved Code of Practice and Guidance.* HMSO, London.

(16)   Flanagan, R. *et al.* (1989) *Life Cycle Costing for Construction.* Surveyors Publications for the RICS, London.

(17)   Ashworth, A. (1999) *Cost Studies of Buildings,* 3rd edn. Longman, London.

(18)   Ashworth, A. & Hogg, K. (2000) *Added Value in Design and Construction.* Pearson Education, Essex.

(19)   Best, R. & de Valence, G. (2002) *Design and Construction – Building in value.* Butterworth Heinemann, Oxford.

(20)   CCPI (1987) *Common Arrangement of Work Sections.* CCPI, London.

(21)   CPIC (2003) *Product Information: a code of procedures for the construction industry.* RIBA, London.

(22)   Searle, M. (1992) 'Project Information for the Construction Industry – is Co-ordinated Project Information the Answer?' Undergraduate dissertation, Leicester Polytechnic.

(23)   Rooley, R. H. (1993) Building Services: Maintenance Systems and Policies. *Structural Survey,* 11:3.

(24)   CIOB (1990) *Maintenance Management – a guide to good practice.* CIOB, London.

(25)   Rooley, R. H. (1992) Building Services: Maintenance Systems and Policies. *Structural Survey,* 11:3.

(26)   Stavely, H. S. (1989) Maintenance management – the consultant's viewpoint, *Chartered Builder,* Nov/Dec.

(27)   www.constructingexcellence.org.uk

(28)   www.dti.gov.uk

Chapter 5
The Nature of Maintenance Work

Much maintenance work will be inevitable, as it is in the nature of materials to deteriorate over time with usage and exposure to the elements of climate. However, the rate at which the deterioration of materials and components takes place may, to some degree, be controlled by prudent decisions being made during the design stage of the procurement process.

Previous chapters have underlined the importance of having properly defined performance standards against which a building's actual performance can be measured, and considerable development has taken place in the methods available for carrying out condition surveys of existing buildings. A well-developed professional building surveying expertise now exists to interpret condition data, often by reference to an extensive body of knowledge on the performance of materials and components, in order to formulate and execute appropriate repair and maintenance strategies.

Despite the presence of this expertise, many property owners do not take professional advice about their buildings. For example, whilst the owner of a factory may investigate the efficiency and energy consumption of the manufacturing plant and machinery, he will rarely require a similar exercise to be carried out for the building fabric, as this is not perceived as contributing directly to productive output.

This means that, whilst it is quite feasible to create a condition model of a building, few building owners avail themselves of the advantages that such a model could bring in devising programmes of repair and maintenance. This short-sighted attitude results in inefficient buildings and potentially higher running costs in the longer term.

Routine maintenance

Routine repairs and remedial action

Routine maintenance, in its purest form, can be thought of as being work that has to be carried out at intervals in order to keep the building in an appropriate condition. This work may involve either the repair or replacement of an item, and is generally necessitated by natural deterioration or normal wear and tear, caused by a wide range of agencies. Common examples of routine maintenance are quinquennial external redecoration and the annual servicing of boilers.

The question to be addressed here is not whether maintenance should be carried out, but rather 'what' and 'when'. For many routine operations there is sufficient data and/or experience to determine a reasonable cycle, and a carefully prepared maintenance manual will supply this information. Once this has been defined, budgets can be framed accordingly and the work executed on a planned cyclical basis.

However, there are factors that militate against this ideal. For example, the correct advice may not be available at hand-over so that:

❏   All routine maintenance items have not been properly identified.
❏   All routine maintenance items have been identified but the wrong advice has been given on the timing and nature of the work required.

On the other hand, all the correct advice may be available but the building owner may choose not to take it. For example, he may perceive that increasing the length of maintenance cycles is a way of reducing overheads in the short term, and use this strategy to either give a market edge or increase profitability. In the public sector it may simply be part of a cost-cutting exercise. The inevitable result is that items that should have been routine become non-routine through postponement to some indeterminate time in the future. Furthermore, there is a multiplier effect, in that failure to carry out routine maintenance at the right time is one of the factors that contributes to abnormal deterioration.

Another reason for not carrying out timely maintenance stems from a lack of knowledge rather than a failure to provide or take notice of information. A design team may, quite conscientiously, produce maintenance guidelines, based on the information that is available. However, in general there is a deficiency in data availability, in terms of both quality and quantity. Whilst it may be unrealistic to expect these information needs to be always met from within an organisation, it might reasonably be hoped that data generated nationally should be more helpful. This was recognised by the Audit Commission, whilst carrying out its 1986 English House Condition Survey,[1] and they commissioned a report from NBA Construction Consultants, which was published ahead of the main report in June 1985.[2] It summarises published knowledge and opinion, derived from 7000 titles and summaries. The authors of the review, however, comment in their introduction that:

> The details of maintenance cycles have to be set properly within the context of the maintenance objectives of individual organisations and the policies resulting from these, particularly the balance between planned and responsive maintenance.
>
> The life expectancy data given here is at best indicative, in many cases representing averages from experience.
>
> In any individual case, element lives will be dependent on many factors, such as design, quality of construction, climatic conditions or degree of use.

Guidance material continues to be produced, and in this respect the publications of Building Maintenance Information (BMI) are recommended as of particular utility.

Replacement of components

Initially, it is convenient to focus on replacements that might reasonably be expected, rather than those that may be termed avoidable. For example, there are items that are replaced on a routine basis as a preventive measure: that is, it is accepted that the item has a finite life and it is replaced before this is reached. This may be a short-, medium- or long-term operation, although it will most commonly relate to items that have a limited life and are therefore replaced, rather than repaired or subjected to routine maintenance. However, even items with a medium- or long-term life may ultimately need to be replaced, having been repaired several times during their life and/or subjected to routine maintenance.

It might be expected that routine replacement items are identified at the outset, together with recommendations for replacement cycles. However, as well as the problem of suitably structured data availability, there are a number of other issues to resolve in determining whether or not to carry out a replacement. This applies, also, to the determination of repair cycles, and both situations are appropriate for the application of optimisation techniques. These issues are explored more fully in Chapter 7.

Planned and unplanned routine maintenance

Under ideal conditions, at the outset of a building's life, it would be possible to have the maintenance of a building completely planned into a series of routine maintenance and replacement cycles. In practice, routine maintenance and replacement may not happen because of failure by the design team to communicate the necessary information to the building owner. Even where such information is provided, the owner may lack the will to use it, or choose not to because of uncertainty.

Despite a professional approach by all parties, unexpected failures of components will occur, and deterioration rates of materials or components may differ from those predicted. Whilst in many instances failure could reasonably have been foreseen and avoided, there still remains a high degree of uncertainty in the prediction process.

Published data will normally present only average values of deterioration rates or replacement cycles. Because it is not possible to be perfectly predictive, continuous monitoring of the performance of a building is necessary in order to update knowledge and improve the reliability of the data already in existence. Routine maintenance programmes must therefore make provision for a system of regular inspection and reporting.

Planned routine inspections

Inspections may be categorised as:

❏   routine inspections that will, perhaps, focus on a predetermined selection of items
❏   major comprehensive inspections, probably on a larger temporal cycle, that consider the whole building or sections of it, and may take a variety of forms depending on the reasons for which they are carried out.

Routine inspections are normally carried out in order to identify items in need of repair or replacement, many of which, being unpredictable, are not readily programmed into a routine maintenance cycle. The determination of appropriate inspection cycles is, of course, no easier than predicting maintenance cycles or replacement periods, and is just as vulnerable to a lack of commitment by the building owner.

Whilst many inspection items, particularly those relating to machinery and plant, may be clearly prescribed, fabric maintenance is rarely the subject of precisely defined inspection cycles. It is to be hoped that, increasingly, inspection checklists will be provided for all the elements of a building, and not just for mechanical plant, as they provide the main means of updating condition data.

By using carefully designed checklists that clearly set out the criteria for classifying the condition of particular elements, it is feasible, after some initial training, to use non-technically qualified staff to collect condition data. Such checklists should incorporate a simple coding system to assist recording, help consistency and indicate when any necessary work should be carried out. In addition to recording items for inclusion in a repair programme, the checklist should provide, through the coding system, a means by which the inspector can 'flag up' any items beyond his/her competence to deal with, for a qualified professional's inspection and advice.

There is often scope for readily incorporating some of the inspection work into day-to-day maintenance operations. For example, when decoration is being carried out, part of the task allotted could be to inspect and report on rainwater goods. If use is made of this procedure, then some care must be taken to ensure that the reporting system employed facilitates the correct flow of information. In other cases inspection work may be carried out by supervisors as part of their general supervisory tours.

Properly designed inspection pro formas should state, using a coding system, who is competent to carry out a particular inspection item, and the person actually executing the inspection should identify themselves, so that reports can also be sorted on this basis. This is important, as there are obviously cases where a specialist inspector is essential.

A systematic approach should also make it possible for the inspector to plan a method and sequence of inspection. Carefully pre-planned pro formas, similar to those described below for condition surveys, help to standardise these procedures, and also lend themselves to automated methods, which are increasingly becoming the norm for managing maintenance data.

If the data is to be imported into a database, then the structuring of the pro forma has to be consistent with the structure of the database being used. All of the automated survey techniques outlined later may be used, and many of the comments made with respect to their relative merits will apply here.

Large estates and buildings also require a methodology for locating items within the building, and the building within the estate. This is normally achieved by means of some form of coding.

To facilitate retrieval, collation and analysis, the facilities and elements inspected should be capable of being grouped by all or any combination of the following:

❑ by location, of either the building or parts of buildings
❑ by elements, such as external walls or windows
❑ by result of the inspection, so that items for urgent repair may be identified easily and quickly.

This clearly identifies the potential for the use of electronic databases and some considerable progress has been made in this respect.

Remedial maintenance

Technical failure

Failure to meet an acceptable standard may be as a result of the normal agencies of deterioration, but equally may be attributable to technical or managerial shortcomings, and the work necessary to restore the standard in such circumstances is classed as remedial rather than routine.

A number of factors may contribute to a technical failure:

❑ inadequate brief collection and development
❑ poor detailed design
❑ buildability problems
❑ poor workmanship
❑ lack of consideration for the execution of maintenance in the design
❑ an inadequate delivery/hand-over package
❑ abuse of the building, in either its use or its management

A substantial amount of time may be spent analysing maintenance and trying to apportion responsibility for a poor performance. Such studies help to provide essential feedback data, but will only be really productive if they are considered to be an integral part of the building design and the process of producing the building. If performance data is to be accurate, it is important to trace failures back to their source, as might reasonably be expected when maintenance work is identified through a comprehensive building survey.

Design and construction failures

These may be considered in two distinct categories. Firstly, failures that occur soon after completion, termed 'burn-in' failures. These are frequently accepted as an inevitable part of running-in the building, and are largely attributable.

Secondly, there are items that fail over a longer period of time, and it is not always possible, or realistic, to properly identify their source. These may be termed 'useful life' or 'wear-out' failures, although they cannot always be considered to be 'normal'.

Between these two will be ongoing failures, which can be classed as 'in-service'.

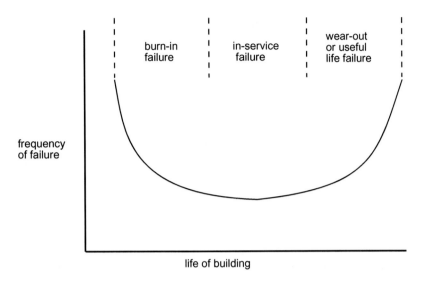

Figure 5.1 Bathtub frequency distribution curve for component failures.

In statistical terms, an analysis of failures produces a frequency distribution called a bathtub curve (figure 5.1), and this may be useful for planning purposes.

However, overlying any consideration of construction failures is the issue of defining precisely what constitutes a failure. Useful life or wear-out failures may often be subtly related to poor decisions made with respect to the choice of components or finishes. These failures are rarely recognised as such, and therefore create deficiencies in the accuracy of feedback data. A large number of failures are thus heavily dependent on perceptions. For example, if an item deteriorates at a more rapid rate than is reasonably expected, it may be classed as a failure by the informed analyst, even though it may not be perceived as such by the user of the building. Many of these failures present further problems, in that they may be latent and not manifest themselves for a number of years, by which time they may have had an adverse effect on some other component or element. This underlines the wisdom of an effective planned inspection policy to detect potential failures at an early stage.

Poor maintenance

It is possible to attribute a significant number of construction failures to poor maintenance practices, such as:

- ❏ inadequate routine maintenance
- ❏ an ineffective replacement programme
- ❏ lack of proper inspections on a planned basis
- ❏ inadequate data to enable the preceding items to be properly carried out.

Classification of maintenance

A classification of maintenance simply into routine or remedial, or planned and unplanned categories is clearly of rather limited value. The Audit Commission considered a better division of maintenance to be the following.[3]

(1)   Strategic repairs and maintenance
This represents work required for the long-term preservation of an asset, and includes planned maintenance of the building fabric (decoration and routine replacement), maintenance of engineering services installations and major repair items such as re-roofing. These are normally items that can be planned for because, to some extent, they can be foreseen and budgeted for.

(2)   Tactical repairs and maintenance
These items relate to day-to-day work of a minor nature, in response to immediate need. The Audit Commission point out that 'tactical maintenance' is not necessarily the same as responsive maintenance, as some immediate response items are clearly of a strategic nature – for example, a flat roof failure.

BS 3811 definitions

The following definitions are all given in BS 3811[4] and, for practical purposes, it is clear that the maintenance workload will consist of a mix of all of these (figure 5.2).

(1)   *planned maintenance* – maintenance organised and carried out with forethought, control and the use of records, to a predetermined plan
(2)   *unplanned maintenance* – ad hoc maintenance carried out to no predetermined plan
(3)   *preventive maintenance* – maintenance carried out at predetermined intervals, or corresponding to prescribed criteria, and intended to reduce the probability of failure, or the performance degradation of an item
(4)   *corrective maintenance* – maintenance carried out after a failure has occurred, and intended to restore an item to a state in which it can perform its required function
(5)   *emergency maintenance* – maintenance that it is necessary to put in hand immediately to avoid serious consequences
(6)   *condition-based maintenance* – preventive maintenance initiated as a result of knowledge of the condition of an item from routine or continuous monitoring
(7)   *scheduled maintenance* – preventive maintenance carried out to a predetermined interval of time, number of operations, mileage, etc.

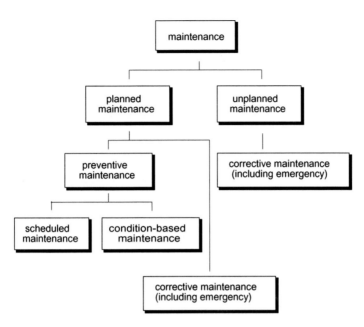

Figure 5.2a Types of maintenance.

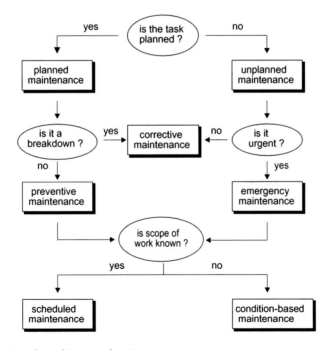

Figure 5.2b Decision-based types of maintenance.

Estate records

It is not possible to manage a property portfolio effectively without an accurate picture of what the estate comprises, the characteristics of the properties and their condition. How comprehensive such information needs to be will be determined by management imperatives, and attitudes to maintenance. A prime requirement, however, is always to have an accurate global property record.

An estate record will consist, in its simplest form, of a list of all the properties, their geographical location and a description of their general characteristics. These characteristics should include occupancy information, technical information concerning the construction of the building and, hopefully, a property history that is accurate and up to date.

Typically, the following might be included:

- ❑ the location of the building
- ❑ landscaping and external works
- ❑ general construction information
- ❑ age and broad condition description
- ❑ services information
- ❑ area and volume of the building
- ❑ type of accommodation
- ❑ type of usage, current occupancy, etc.
- ❑ a brief history of the building including any alterations or additions that have been carried out and when
- ❑ any proposals for future use or alterations
- ❑ current value of the property.

Such a property record will then form part of the estate or portfolio record, within which it may be given a reference number or code. This information can normally be obtained quite readily, stored in an accessible form and, with good management, be easily kept up to date.

At some stage a detailed condition survey will be required for each property, necessitating a full rather than a simple inspection.

Condition surveys

The common local authority practice of estimating maintenance expenditure by simply taking a notional percentage of the property value has been heavily criticised, and the use of proper condition surveys to derive more accurate estimates of maintenance expenditure is strongly recommended. In defence of the practice, the full cost of such surveys has to be recognised, along with issues with respect to 'shelf-life'. Condition surveys should, however, be commissioned for more than just budgeting purposes, as they have a wider application in the managing of building condition.

A major obstacle to carrying out the first comprehensive survey is the expense. On a national scale the UK building stock possesses very poor condition records, and this represents a massive impediment to developing good maintenance management practices. Some progress has been made in recent years, particularly with respect to local authority buildings, where the prompting of the Audit Commission has had some effect. Within the private sector, however, there is still a startling reluctance amongst property managers to commit funding and commission detailed condition surveys of their buildings.

Even within the public sector, many of the condition surveys now being carried out are strictly limited in their scope. In many cases they are carried out for very specific purposes, usually related to financial management, rather than as part of a professional approach to managing building condition. This is very wasteful of scarce resources, as instead of one comprehensive condition survey being carried out, which could be kept up to date by periodic inspection, several ad hoc surveys are undertaken, the data from which is impossible to unify into a comprehensive picture.

Building surveys

Due to some confusion amongst clients as to the precise meaning of the various forms of building survey carried out, the Construction Industry Council and a range of contributing bodies have produced a set of clear definitions. It is of note that the term 'structural survey' should not now be used.

The following types of survey are defined and included here for clarity:

(1) Valuation
This is primarily to provide an opinion on the price a property might achieve if sold or rented in the marketplace. It is thus a simple exercise in technical terms limited to a view of constructional features from a walk-around inspection.

(2) Property purchase survey and valuation
This type of survey, as well as providing a valuation, will provide factual information on important aspects of the building condition. It is of particular use for relatively simple properties, e.g. domestic. Where the building is more complex, larger, older or of non-standard construction, a full building survey is recommended.

(3) Building survey
This is defined as:

> . . . an investigation and assessment of the construction and condition of a building and will not normally include advice on value.

Such surveys may recommend that more detailed elemental or specialist investigations are carried out. These are indepth studies but more limited in scope.

(4) Stock condition survey
This survey is commissioned to assess the condition of an estate or part of it to enable a planned maintenance programme to be developed.

(5)   Schedule of condition

This records the condition of a building at a point in time and will probable include documentary evidence such as dated photographs, sketches and drawings, typically at the beginning of a lease.

(6)   Schedule of dilapidations

This is carried out to identify the needs of tenanted properties in the context of repairing obligations under the terms of a lease.

In planning and carrying out any survey, the three following interrelated issues need to be considered:

❑   survey procedures and collection of data
❑   presentation and structuring of data
❑   data storage

There is a great deal of difference between carrying out a one-off survey and surveying a large portfolio of buildings. When preparing a property database for maintenance management purposes, data storage will be an important and potentially expensive factor. This has prompted increasing interest in cost-reducing survey techniques, which at the same time collect data in a form compatible with some overall management information system.

At the simple level, use of a pro forma type survey is obviously attractive, as it enables a degree of standardisation. However, this approach requires very careful planning and consideration of the requirements at the outset, and is only realistic if there is the possibility of using the pro forma a number of times.

Databases used for the storage and manipulation of large quantities of data are ideal for planning purposes, and provide an excellent tool for analysis which, if properly integrated within the management system, can be of great benefit in facilitating the transfer of information. If a database is to be used, important issues have to be resolved with respect to the structuring of the information. Building condition data for a group of houses lends itself rather easily to a database structure based simply on construction elements, all of which repeat themselves a number of times. In a more heterogeneous property portfolio, though, the question of the database structure is by no means a simple one.

The nature of the database structure, once decided on, will determine survey procedures and the preparation of pro formas. Although the principles for traditional manual surveys have been firmly established over the years, there are still many schools of thought with respect to the exact procedures to be followed, and the automation of any part of the process seriously calls into question the appropriateness of these traditional approaches. Developments in electronic data collection and storage require surveyors to re-appraise their methods at all levels.

Whatever approach is taken, the principal aim of a building condition survey is to provide an accurate picture of the property stock. In general the condition survey requires:

❑   an accurate description of each building element, which describes both the construction and the materials used
❑   an assessment of the state of repair for each element, together with a statement of obvious defects
❑   an indication of the expected physical life of each element
❑   a schedule of repair work required, and an indication of the priority assignable to each item
❑   a cost estimate of the work required, which may not be provided at the time of the actual survey, but as a follow-up exercise.

The amount of detail and depth of coverage required for each element under consideration is extremely variable, and in practice will be determined by the precise requirements and purposes of the survey.

In terms of data collection, three approaches can be identified:

❑   manual
❑   optical mark and bar-code readers
❑   hand-held computer.

Manual methods

Manual surveys are frequently conducted using a pro forma, which acts as a guide to the information to be collected. This may range from a checklist or series of prompts, to a rigidly structured survey form. The former may give more flexibility to the surveyor in terms of his survey technique, but this is often offset by the need to restructure the information to suit data storage needs. Whilst this approach lends itself very well to the collection of factual material, it cannot be a substitute for the judgemental aspects of the surveyor's work, such as assessing condition and useful life of a component.

Many of the surveys currently being undertaken by local authorities for maintenance management purposes are very objective, in that they are highly structured rather than judgemental. This is a result, not only of financial constraints, but also of a limited set of survey objectives (figure 5.3).

The surveyor's knowledge and experience will always be vital; a systematised approach can never give a full picture, and will certainly be weak diagnostically. Notwithstanding this, the major problem encountered with the manual approach is the difficulty of maintaining consistency between surveys and the surveyors carrying them out.

In considering the production of planned maintenance programmes, consistency is necessary and this may be achieved by using more sophisticated methods.

Optical mark and bar-code readers

The optical mark reader (OMR) uses a pre-printed form, designed specifically for the task in hand, which consists of a series of multiple-choice questions. The reply to each

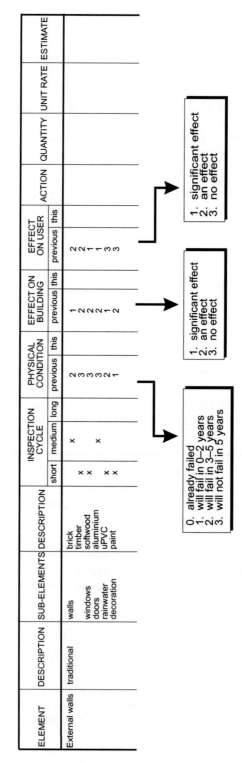

Figure 5.3 Example of condition survey pro forma.

question is marked in the space provided next to the appropriate option, using a graphite pencil. When the form is completed, the survey sheet is fed into an OMR for processing.

The survey sheets are printed in a colour that the computer-based reader is programmed to ignore. It therefore only reads the graphite pencil marks, interpreting them according to their position on the sheet. In most applications the information is stored in a database, which may also contain a built-in schedule of rates, permitting rapid assessment of cost implications (figure 5.4).

The bar-code reader uses pre-printed multiple-choice forms, as described above, but instead of a mark, each question is allocated a bar code. As the survey is carried out the surveyor uses a light pen against the bar code for the appropriate answer. The light pen is connected to a hand-held portable computer, which stores the data in a database format as the survey proceeds. On completion of the survey the portable computer downloads the data to the main computer in the office.

The merits of this technique are very similar to those of the previous one, and the major advantage is in the reduction of labour at the data transfer stage. On the other hand, the equipment used is more cumbersome.

Difficulties can arise with both approaches when so called 'rogue' items, which are not included in the schedule, are observed. These will need to be priced manually, which can be very disruptive when the whole work environment is electronically based. Programmes can be written that provide built-in checks for consistency and gross errors, but the systems are limited by the extent to which contingencies are provided for, both on the form and in the software. The two have to be compatible, and hence there may be a tendency to lose accuracy in the interests of expediency.

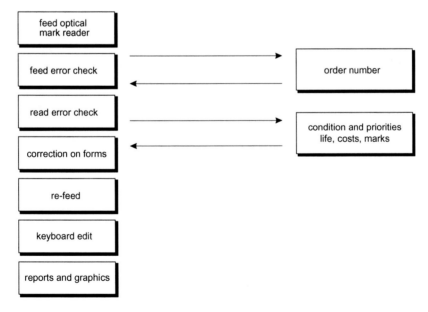

Figure 5.4 Procedure for feeding survey forms to OMR.

The consistency of the information gathered is, however, improved drastically and they are an excellent analysis tool. It could be argued that this has more to do with the use of a structured survey sheet than the techniques *per se*. Their application for carrying out stock condition surveys is today rather limited but for asset management and tracking they have much to commend them.

Hand-held computer

There has been rapid development of hand-held computers and many commercially available products now provide a fully integrated tool. In a simple form a hand-held computer can be programmed, prior to the survey, with a series of questions that can be multiple choice, or devised to act as prompts. The surveyor answers the questions as the computer asks them, so the order in which they are asked is crucial. Built-in checks can be programmed which, being operative during the survey, provide a major advantage. These checks may include a requirement that no question can be missed, and range checks that prohibit the entry of absurd answers. After the survey the hand-held computer can be downloaded as before, and figure 5.5 indicates a typical process.

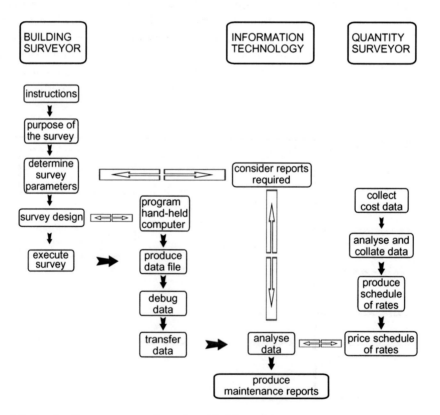

Figure 5.5 Conducting a survey using a hand-held computer.

More sophisticated tools now available form part of an automated planned maintenance package, with the computer having a direct link with either bespoke or tailored
databases.

The major advantage of using hand-held computers, compared to pre-printed pro
formas, is the flexibility they provide in using standard software that can be customised to suit the needs of each survey.

For asset-tracking, combinations of bar-code readers and hand-held computers are
available rather like the stock-management systems employed by large retailers. This
approach may also be used for stock condition surveys (figure 5.6).

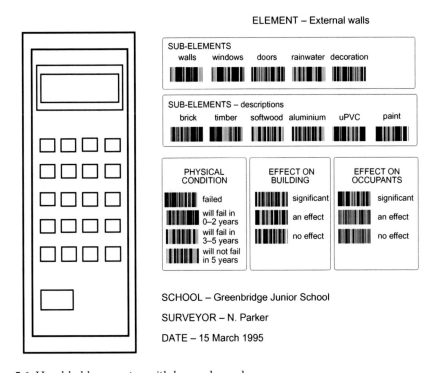

Figure 5.6 Hand-held computer with bar-code reader.

Reports

The information collected from a building survey can be presented in a number of
ways. It can be in the form of a written report, in a spreadsheet format or in the form
of charts and graphs (figures 5.7 and 5.8). The first of these is appropriate to individual
bespoke surveys, but for maintenance management purposes, charts, schedules and
graphs are more likely to be appropriate.

What will have been obtained in all cases is a picture of the condition of the
buildings making up the estate. The major cost is in executing the first survey; in

BIENNIAL SCHOOL SURVEY: Summary of findings on elemental basis

SCHOOL – Greenbridge Junior School

SURVEYOR – N. Parker

DATE – 15 March 1995

ELEMENT	SUB-ELEMENTS	GENERAL CONDITION	ACTION	EST. EXPEND. (£)					FUTURE ACTION
				year 1	year 2	year 3	year 4	year 5	
Substructure	strip foundations concrete slab blue brick DPC	all good (visual inspection only)							
External walls	brickwork timber cladding	good fair	decorate and monitor	400	50	50	500		possible replace
Windows and external doors	softwood aluminium	fair good	decorate and monitor	500	50	50	500		possible replace
Roofs	traditional – pitched flat – felt rainwater	good fair fair	renew clean out	75	75	350 75	100	100	
Internal fabric	blockwork stud partitions	good fair	repair	200					
Internal finishes	walls floors ceilings paintwork	good fair good fair	replace carpet decorate		800 750				possible repairs
Services	gas water electricity internal drainage fittings	good (visual inspection) ditto ditto ditto fair	detailed inspection ditto ditto ditto replace			50 100 50 100 250	250		possible future work on all services following inspection
External works	paths steps walls external drainage roads and car parks landscaping	good fair fair fair fair good	repair repair clean resurface	300 100	400 100	100 1000	200 200	200	

Figure 5.7 Summary of condition survey recommendations.

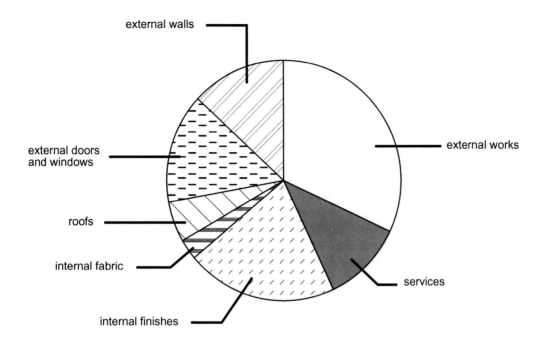

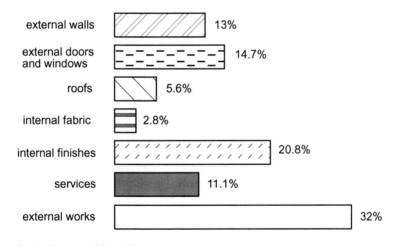

Figure 5.8 Use of pie charts and bar diagrams.

many instances the picture obtained may be limited to that required by immediate demands, and thus it may often only provide sufficient information for budgeting purposes.

Statutory requirements

The legislative framework for building standards is principally concerned with matters affecting health and safety, and is set out in Acts of Parliament, which are supplemented by detailed regulations that are published from time to time. The Acts range from those that relate to buildings in general to those that cover the condition of specific types of building.

The legislative provisions can be divided into those that control the design and physical requirements of new buildings, alterations and extensions, such as Town and Country Planning Acts and the Building Act, and those that relate to the occupancy of a building. The latter mainly come under the provisions of the Health and Safety at Work etc. Act.

Planning Acts

The major principle of these is to control the use to which land is put, and they cover not only new buildings but changes to existing buildings, in either physical characteristics or use. These Acts can then, in some circumstances, have relevance to the execution of major repair work, where planning requirements may control the nature and appearance of any work carried out.

Of greater importance under this heading, however, are the powers given to the Secretary of State to compile lists of all buildings that are of historical or architectural merit. Listed building consent must be obtained from the local planning authority before any work is carried out on such a building.

Thus far, planning can be seen as a controlling influence on what is carried out and how it is done, rather than a generator of work. However, it is worth noting that a local authority has the power to prevent an owner from allowing a listed building to deteriorate, by serving him with a repairs notice specifying the work required to preserve the building.

The Secretary of State also has powers to make grants for the maintenance and repair of buildings of 'outstanding' architectural or historic interest, and for the preservation or enhancement of character or appearance of conservation areas.

Building Act 1984

This Act consolidated much previous primary legislation relating to the physical requirements of buildings, including fundamental changes in the method of building control. This is essentially enabling legislation under which subsequent controls can be extended and/or modified, and there are several parts.

(1)   Part I is concerned with Building Regulations. Under section 5.1 the Secretary of State is empowered to make regulations with respect to the design and construction of buildings and the provision of services, fittings and equipment, for the purposes of:
  ❏   securing the health, safety, welfare and convenience of occupants and other people affected by the building
  ❏   furthering the conservation of power and fuel
  ❏   preventing waste, undue consumption, misuse and contamination of water.

  Section 5.2 is of particular importance, in that it provides that Building Regulations may impose continuing requirements on the owners and occupiers of buildings, including those that were not, at their time of erection, subject to regulation. These may apply to what are termed 'designated functions', such as keeping fire escapes clear; or, in respect of some services, fittings and equipment, such as the periodic inspection and maintenance of lifts.

(2)   Part II relates to the arrangements for supervision of work other than by local authorities.

(3)   Part III gives local authorities the power to require the owners or occupiers of buildings to provide certain facilities, or remedy certain deficiencies, and it may serve notices requiring work in connection with the following typical items:
  ❏   drainage outside the building
  ❏   sanitary provisions in a building used as a workplace
  ❏   quality of sanitary provision
  ❏   means of escape in case of fire, in consultation with the fire authority
  ❏   raising of chimneys
  ❏   alteration or closing of cellars and rooms below subsurface water levels.

  It may also serve notice on an owner or occupier that it intends to carry out certain actions itself, for example:
  ❏   where a building is in such a state, or is used to carry such loads, as to be dangerous
  ❏   the repair, restoration or demolition of a building which is, by its ruinous or dilapidated condition, seriously detrimental to the amenities of an area
  ❏   paving and drainage of yards and passages.

  The local authority may apply to a magistrate's court for an order to:
  ❏   temporarily close or restrict the use of a building, pending the provision of satisfactory means of ingress and egress
  ❏   execute remedial work to a dangerous building or restrict its use, where danger arises from overloading.

In some cases, persons carrying out certain types of work are required to give the authority a prescribed period of notice before starting the work. This occurs, for example, in the case of altering the course of an underground drain.

Building regulations

These apply to building work in England and Wales, and to certain changes of use of an existing building. Building work is defined as:

- ❏ the erection or extension of a building
- ❏ the material alteration of a building
- ❏ the provision, extension or material alteration of a controlled service or fitting
- ❏ work required on a material change of use
- ❏ insertion of cavity wall insulation
- ❏ work involved in the underpinning of a building.

The regulations say nothing specifically about the point at which repair becomes subject to control. Normally this would not be controlled, but if the point is reached where a building has been seriously damaged or dilapidated, the authority may reasonably apply the regulations to any remedial work.

In general the Building Regulations will be relevant only to the nature of a repair, or the way in which it is carried out, rather than a generator of maintenance *per se*. However, compliance with the provisions of the regulations will in many circumstances lead to improvements and impinge on maintenance in the broadest understanding of the term – for example, replacement windows are required to be double glazed.

Housing Acts

These are of prime importance when maintaining housing stock. Indeed, in many situations, where resources are scarce, they lay down the minimum standard that a housing authority must provide, and become the major maintenance planning yardstick.

The Housing Act 1957, re-enacted in 1985, laid down criteria for determining whether or not a house is unfit for human habitation. Subsequently the 2004 Housing Act (Part 1) introduced the notion of a Housing Health and Safety Rating System, as outlined in Chapter 2.

Defective Premises Act 1972

The purpose of this Act is to impose duties in connection with the provision of dwellings, and to attribute the liability for injury or damage caused to persons through defects in the state of premises. The Act has potentially far-reaching implications for landlords and building owners.

In the first instance, the Act imposes a strict duty to build properly. This duty is owed to the person for whom the dwelling is provided, and to every person who subsequently acquires an interest in it. A person who does work to any premises is under a duty at common law to take reasonable care for the safety of others who might reasonably be expected to be affected by defects arising from their work.

Where premises are let under a tenancy that obliges the landlord to maintain and repair the building, the landlord owes to all persons who might reasonably be expected to be affected by a defect a duty to see that they are reasonably safe from personal injury and from damage to their property caused by the relevant defect. This duty is owed by the landlord if he knows, or if he ought in all circumstances to have known, of the relevant defect. There are clear implications here in terms of repair and maintenance for the landlord, and a direct encouragement to him to carry out regular inspections of his properties.

Health and Safety at Work etc. Act 1974 (HASWA)

This Act provides a comprehensive and integrated approach, which governs the safety, health and welfare of people at work, and members of the public who may be affected by such work.

The Act is an enabling Act and has provision for the development of a framework of health and safety statutory instruments – regulations – and any associated standards and approved codes of practice (ACOPs). Whilst standards and codes of practice are not legally binding, compliance with them is admissible to a court of law as evidence of satisfying the requirements of regulations.

The provisions of HASWA are to:

❏ secure the health, safety and welfare of persons at work
❏ protect persons at work against risks to health or safety arising out of, or in connection with, the activities of persons at work
❏ control the keeping and use of explosive, highly flammable or dangerous substances
❏ control the emission of noxious or offensive substances into the atmosphere.

It is the duty of every employer to ensure, so far as is reasonably practicable, the health, safety and welfare of all his employees, and in addition to this general duty, particularly to ensure:

❏ the provision and maintenance of plant and systems of work that are safe and without risks to health
❏ that arrangements are made to remove risks in connection with the use, handling, storage and transport of articles and substances
❏ the provision of information, instruction, training and supervision
❏ that any place of work is maintained in a safe condition without risks to health, and the provision of safe access and egress
❏ the provision of a safe, risk-free working environment, with adequate facilities and welfare arrangements.

Except as prescribed, every employer must prepare and update a written health and safety policy and the organisation and arrangements for carrying it out, and bring it to the notice of all employees.

It is the duty of every employer and self-employed person to conduct their undertaking in such away as to ensure that persons not in their employment are not exposed to risks to their health and safety.

The Health & Safety Commission (HSC) and Health & Safety Executive (HSE)

The Act creates and confers powers upon the HSC and HSE. The HSC is a corporate body which serves to:

- assist and encourage matters relevant to the promotion of health and safety
- undertake and publish research
- provide an information and advisory service
- submit proposals from time to time, regarding the making of relevant regulations
- give effect to any directions given to it by the Secretary of State.

The HSE is a corporate body which serves to:

- undertake the Commission's functions as directed by the Commission
- to give effect to such directions given by the Commission

The HSE is in effect the operational arm of the HSC.

The Secretary of State is given powers to make, repeal, modify and adapt health and safety regulations and ACOPs.

In 1992 in response to the following EU Directives:

- The Workplace Directive 89/654/EEC
- The Use of Work Equipment Directive 89/655/EEC
- The Personal Protective Equipment Directive 89/656/EEC
- The Manual Handling of Tools Directive 90/269/EEC
- The Display Screen Equipment Directive 90/270/EEC
- The Construction Sites Directive 92/57/EEC,

HSE introduced the following regulations:

- *The Workplace (Health, Safety and Welfare) Regulations 1992* – otherwise known as 'the Workplace Regulations'
- *The Provision and Use of Work Equipment Regulations 1992* (replaced by 1998 Regulations) – otherwise known as 'the Work Equipment Regulations (PUWER)'
- *The Personal Protective Equipment at Work Regulations 1992* – otherwise known as 'the Personal Protective Equipment (PPE) Regulations'
- *The Manual Handling Operations Regulations 1992* – otherwise known as 'the Manual Handling Regulations'
- *The Health and Safety (Display Screen Equipment) Regulations 1992* – otherwise known as 'the VDU Regulations'
- *The Construction (Design and Management) Regulations 1994* – otherwise known as 'the CDM Regulations'.

The maintenance manager may have much of the responsibility for protecting the interests of the occupier in providing safe and healthy working conditions, as the Act places a responsibility on the person in control of the premises to ensure that not only are the premises themselves safe, but also the plant and machinery therein. Thus, the maintenance manager, depending on his defined role, could incur personal liability for non-compliance.

The maintenance manager must therefore consider himself as being not only in a position where he needs to respond to requests for work to remedy a defective situation, but also one in which he must proceed in a proactive way, so that compliance with the legislation becomes part of a programme of planned inspection and maintenance.

Management of Health & Safety at Work Regulations 1999 (MHSWR)

These regulations give effect to the European Framework Directive on health and safety. They supplement HASWA and specify a range of management issues, most of which must be carried out in the workplace. An ACOP, *Management of Health and Safety at Work. Approved Code of Practice* (HSE 2000), supports the regulations.

The MHSWR require employers to undertake the following:

- carry out risk assessment to determine the risks to health and safety of their employees and others
- make arrangements to implement preventive or protective measures to mitigate the risks identified
- monitor activities for the occurrence of risks
- appoint competent persons to *manage* risk
- provide health and safety information to employees (and to temporary and visiting workers).

Risk assessment is an important aspect of implementing regulations.

Risk assessment

Risk assessment is a well-established and recognised analytical technique widely used across many different fields of business, commerce and industry. It involves three key processes:

- hazard identification
- evaluation of risk
- prevention and protection measures.

(1) Hazard identification
A *hazard* is anything that has a potential to do harm. This could be through the occurrence of an accident or exposure to a dangerous situation, material or substance.

Ever-present hazards in the construction industry that lead to fatal and serious injury commonly involve:

❑   working at heights
❑   use of ladders
❑   collapse of temporary structures
❑   use of vehicles, plant and equipment
❑   exposure to harmful substances.

(2)   Evaluation of risk

Risk is the likelihood that a specified event will occur as a result of the realisation of a hazard. Once a hazard has been identified, the *degree of risk* must be determined. Two factors are influential to this determination:

(a)   the *severity of harm* – the level of harm that a circumstance would create
(b)   the *likelihood of occurrence* – the frequency of the hazardous circumstance.

The evaluation of risk can be made with the aid of simple calculation:

$$\textit{degree of risk} = \text{severity of harm} \times \text{likelihood of occurrence}$$

Using the evaluation criteria in Tables 5.1 and 5.2, the degree of risk can be calculated; for example, in a situation where the severity of harm is 4 and the likelihood of occurrence is 3 the degree of risk will be 12.

A priority can be assigned to the risk, using the following ranges:

❑   1–5   low priority
❑   6–16   medium priority
❑   17–36   high priority

The greater the degree of risk value, the higher the priority; therefore the more thought that should be given to, and the more effort that should be placed upon, avoiding or managing the risk.

Table 5.1 Evaluation criteria for severity of harm.

	Description	Assigned value
LOW	Minor injury	1
	Chronic illness	2
MEDIUM	Acute illness	3
	Reportable injury (as defined by RIDDOR*)	4
HIGH	Major injury (as defined by RIDDOR)	5
	Death	6

*Reporting of Injuries, Diseases and Dangerous Occurrences Regulations 1995

Table 5.2 Evaluation criteria for likelihood of occurrence.

	Description	Assigned value
LOW	Remote – almost certain not to occur	1
	Unlikely – occurrence in exception circumstances	2
MEDIUM	Possible – known occurrences over 12 months	3
	Likely – occurring monthly	4
HIGH	Probable – occurring daily	5
	Highly probable – occurring all the time	6

A typical hierarchy of control, from high risk to low risk, is indicated below:

(1) *elimination* of the risk completely, e.g. prohibiting a certain practice or the use of a certain substance
(2) *substitution* with something less hazardous or risky
(3) *enclosure* of the risk in such a way that access is denied
(4) *fitting guards or installing safety devices* to prevent access to danger points or zones on work equipment and machinery
(5) *safe systems of work* that reduce risk to an acceptable level
(6) *written procedures,* e.g. job safety instructions that are known and understood by those affected
(7) *adequate supervision,* particularly of young or inexperienced persons
(8) *training* of staff to appreciate risks and hazards
(9) *information,* e.g. safety signs, warning notices
(10) *personal protective equipment,* e.g. eye, hand, head and other forms of body protection.

The Workplace (Health, Safety and Welfare) Regulations 1992 (WHSW)

These regulations came into force for all workplaces on 1 January 1996 and are supported by the Workplace (Health, Safety and Welfare) Regulations 1992: Approved Code of Practice and Guidance 1992 (ACOP).

WHSW regulations place a duty on an employer in respect of workplaces under their control. They are concerned with the safety and welfare of employees working in the building. The main provisions relate to:

❑ *Maintenance* – the workplace and associated equipment must be maintained in a clean and efficient state, and be in efficient working order and in good repair.
❑ *Ventilation* – where windows or other openings do not provide suitable ventilation, mechanical ventilation shall be provide and maintained.
❑ *Temperature* – a reasonable temperature should be provided in workrooms. Generally, 16°C would be a minimum temperature. Where work involves significant physical effort, 13°C would be more appropriate.

❑ *Lighting* – natural lighting should be provided where reasonably practicable, therefore windows and roof lights must be kept clean. Lighting should be sufficient to enable people to work and move safely without visual fatigue. Local and emergency lighting should be provided where necessary.

❑ *Cleanliness and tidiness* – floors, walls, ceilings and furnishings must be kept clean, the standard of cleanliness required depending on the use of the workplace. (Some other regulations, e.g. Food Hygiene Regulations, have specific requirements.)

❑ *Workstations and seating* – workstations should be so arranged that each task can be carried out safely and comfortably, provide adequate support for the lower back and a footrest where the foot cannot comfortably be placed flat on the floor.

❑ *Conditions of floors and traffic routes* – floors of workplaces and surfaces of passages, staircases, access roads, etc. must be suitable for their intended use and properly maintained.

❑ *Falls and falling objects* – suitable and effective measures shall be taken to prevent any person falling a distance, or being struck by a falling object, likely to cause personal injury.

❑ *Windows and transparent or translucent doors, gates and walls* – glazing of doors and windows that could be broken accidentally by persons or materials and cause injury must be made of a suitable safety material or be adequately protected against breakage by suitable screens or barriers.

❑ *Windows, skylights and ventilators* – windows and skylights etc. must be capable of being opened and closed without risk.

❑ *Ability to clean windows etc. safely* – windows should be designed, or provision made, to ensure that cleaning can be carried out safely.

❑ *Traffic routes* – traffic routes must allow the safe movement of persons and vehicles within the workplace and when entering or leaving (which might include the segregation of pedestrians from vehicles).

❑ *Doors and gates* – doors and gates should be suitably constructed and those that swing in both directions should have a transparent panel. On main routes, all doors should be fitted with such panels.

❑ *Power-operated doors and gates* – power-operated doors and gates should have appropriate safety features to prevent injury to persons if struck by them.

❑ *Escalators* – escalators and moving walkways must be fitted with necessary safety devices.

❑ *Sanitary conveniences and washing facilities* – suitable and sufficient facilities must be provided for the maximum number of persons likely to be at work in the workplace at any one time.

❑ *Drinking water* – an adequate supply of wholesome drinking water must be provided, with suitable cups unless supplied by a drinking fountain. Such supplies should be clearly marked if there is any risk of people drinking contaminated supplies.

❑ *Accommodation for clothing* – suitable and sufficient accommodation must be provided for any special work clothing and for personal clothing that is not worn at work. Clothing should be hung in a clean, warm, dry place with a hook for each worker.

❑ *Facilities for changing* – where workers are required to wear special work clothing, adequate room for separate-sex changing should be provided and measures should be taken to ensure security, e.g. by providing lockers.

❑ *Facilities for rest and to eat meals* – suitable seats should be provided for workers whose work gives them an opportunity to sit.

Fire safety

At the time of writing the control of fire safety in premises is the subject of a major reform. The aim is to reduce burdens on business that are caused by the existence of multiple, overlapping general fire safety regimes. It will do this by consolidating and rationalising much of the existing fire safety legislation, principally contained in the Fire Precautions Act 1971, the Fire Precautions (Workplace) Regulations 1997, and a plethora of licensing, certification and registration regimes where fire safety is an issue – for example, Licensing Acts, Theatres Act, Gaming Act, and various children's homes regulations.

The Regulatory Reform (Fire Safety) Order 2005 (RRFSO) should not impose significant additional burdens on employers since it recreates the requirements that already exist under:

❑ Fire Precautions (Workplace) Regulations 1997
❑ Management of Health & Safety at Work Regulations
❑ Dangerous Substances & Explosive Atmospheres Regulations 2002,

and replaces them with a simple fire safety regime applying to all workplaces and other non-domestic premises.

Under the RRFSO, Fire Certificates are abolished and will cease to have legal status.

The RRFSO adopts a risk-assessment-based regime, with responsibility for the fire safety of the occupants of the premises and people who might be affected by a fire resting with a defined *responsible person* who could be:

❑ the employer (if there is one)
❑ the person in control of the premises in connection with the carrying on of a trade, a business or undertaking (for profit or not)
❑ the owner
❑ any other person who to any extent exercises control over the place.

The new regime requires the responsible person to ensure that:

❑ an assessment of risks is undertaken
❑ hazards are removed or reduced as far as is reasonable
❑ relevant persons are protected from the hazards that remain
❑ particular attention is paid to those at special risk (disabled people, those with special needs)

❑ all precautions are subject to installation and maintenance by a 'competent person'.

This approach promotes the avoidance of fires and the mitigation of the effects of fire to a level equal to the current position of securing safe escape. It does not detract from the need for precautions to ensure safe escape in case of fire.

The main enforcing body will be the local fire and rescue authority, except in some special cases where it is appropriate for another authority to enforce, for example the HSE for construction sites and the local authority for the issue of sports ground safety certificates.

Enforcing authorities will be able to require work to be carried by *enforcement notices*, and in extreme cases *prohibition notices* can be used to prohibit or restrict the use of premises until such time as they have been made appropriately safe.

Obligations under tenancy agreements

The apportionment of responsibility, between landlord and tenant, to maintain a building subject to a lease, can be provided for by covenants in the lease agreement. In theory this should be quite a straightforward matter, but in practice it can become quite a complex affair. A body of case law on dilapidations has developed over the years and this continues to grow.

The word 'dilapidation', in general use, expresses some notion of disrepair. When used by a lawyer or surveyor, it normally refers to a property that is not only in a state of disrepair, but where, as a consequence, a legal liability is incurred. This liability is normally incurred by virtue of a lease which has placed an obligation on either the tenant or the landlord to maintain and repair certain parts of the property. Liability may also be incurred, however, under the law of tort, or by statute.

Waste

Waste may be defined as an act or omission which causes a lasting alteration to the land or premises affected to the prejudice of the owner or landlord. Waste is a tort and independent as a source of liability from any contractual relationship expressed or implied, between parties, so that an obligation not to commit waste is not excluded by an express or implied covenant in a lease dealing with the same subject matter.[5]

(West's Law of Dilapidations)

There are a number of types of waste:

❑ 'Voluntary waste' is the deliberate carrying out of some act that tends to destroy or injure land, premises or landlord's fixtures.
❑ 'Permissive waste' is allowing buildings to fall into disrepair by neglect, and failure to repair them.

❏   Any unauthorised change in the nature or character of premises is waste; however, if changes result in the betterment of the premises then this is termed 'ameliorating waste'.

Liability for waste is founded in tort, and the measure of damages is the amount of damage to the landlord's reversion – that is, the depreciation in sale value of his interest, which may not be the same as the cost of reinstatement. Actions in waste are uncommon, but may be the only redress against persons in occupancy of premises without a tenancy, or other persons in occupation of land without express or implied obligations to repair.

Statute

Parliament has modified the common law rule of no implied liability on landlords to carry out repairs or see that premises are fit or suitable for occupation by the tenant, under the Landlord and Tenant Act 1985. This applies in the case of short leases of dwelling houses, and imposes certain repairing liabilities on landlords, which may not generally be contracted out of.

Landlord and tenant obligations under leases

The obligations to repair and maintain are normally laid down in repairing covenants. In a general covenant to repair, the tenant will undertake one or all of the following obligations: to put, keep or leave the premises in a defined state of repair.

A specific covenant to put in repair by the tenant would be unusual, since it is implied by a covenant to keep in repair that the tenant will first put the premises into repair if they are in disrepair. Where dilapidated premises are to be let, this type of covenant could be used to get the building put into repair early in the life of the lease. In such a case the covenant will usually specify a particular standard of repair, and timescales will be given for the necessary work.

With a covenant to keep in repair, it may be implied that the tenant will have received the premises in good repair at the commencement of the lease. The condition of the building at this time may be regarded as the benchmark of the standards required, and should therefore be the subject of a schedule of condition, agreed by both parties at the outset.

If the only repairing covenant is to leave the premises in repair, then no action can be brought against the tenant, during the lease, to carry out repairs. The landlord must wait until the end of the lease. This underlines the importance of establishing a schedule of condition at the creation of the lease.

There are a number of problematic issues, all the subject of a large body of case law:

❏   the meaning of repair
❏   standards of repair
❏   fair wear and tear.

This is not the place to comment in detail, as a thorough treatment of these issues is provided by Hollis.[6]

Most leases contain a clause requiring the tenant to be responsible for carrying out works that may be required to comply with current statutes, and there are a number of possible remedies for breach of repair covenants. The applicability of each of the following depends on the nature of the lease and individual circumstances:

❏   forfeiture of the lease, in which case the tenant may also be required to make monetary compensation
❏   injunction to prevent a party to a contract from carrying out some act which he has been covenanted not to do
❏   a court decree compelling a party to carry out what he has been covenanted to do
❏   the landlord executes the repairs, which is only possible if there is an express provision in the lease for him to enter and execute repairs
❏   where a tenant has covenanted to repair and failed to do so, then damages may be recovered, and this amount will be determined by whether or not the lease is still operative or has terminated.

Condition surveys of rented properties are obviously, therefore, of particular importance, at both the beginning and end of a lease, and there may be a specific requirement, depending on circumstances. Schedules of dilapidations may often be prepared, frequently by both parties, with the intention of defining repair obligations, rather than for use as a management tool.

References

(1)   Department of the Environment (1988) *English House Condition Survey 1986.* HMSO, London.
(2)   Construction Consultants (1985) *Maintenance Cycles and Life Expectancies of Building Components: A guide to data and sources.* HMSO, London.
(3)   Audit Commission for Local Authorities in England and Wales (1988) *Local Authority Property – a Management Handbook.* HMSO, London.
(4)   British Standards Institute (1984) *BS 3811: 1984 Glossary of Maintenance Management Terms in Terotechnology.* HMSO, London.
(5)   West, W. A. (1979) *The Law of Dilapidations.* Estates Gazette, London.
(6)   Hollis, M. (1988) *Surveying for Dilapidations.* Estates Gazette, London.

Chapter 6
Information Management

Introduction to information systems

Maintenance operations require, and in turn generate, large amounts of data. The collection and processing of this data into management information is, therefore, a key issue in managing building maintenance. Whether manual or automated means are employed, there exists an information system in one form or another. In this sense information systems should not be confused with IT, as in some organisations they may still in part be manual.

Information management involves the design of a system in which data is collected and processed into management information. This makes several important distinctions in terminology, and emphasises that information is a function of data, a process and the needs of the user.

A system can be represented simply, as shown in figure 6.1, in terms of input and output. Figure 6.2 represents a simple view of an information system related to condition surveys, and illustrates a method of describing the elements of a system.

There are a number of issues that stem from this simple model.

(1) Raw data is subjected to processes that convert it into management information. In some instances this process may generate new data.
(2) Data may be collected and/or generated within the system, or may exist as an external entity.
(3) Information systems exist to serve user needs, which therefore need to be defined. The system must then be designed to produce information appropriate to those needs.

The information system can be viewed as having three essential components, relating to the organisation, data operations and technology. In business terms this can be represented as shown in figure 6.3.

Information systems and the organisation

Organisations' structures grow into existence in a wide variety of ways, to satisfy a number of objectives. Information management will be one of these but not until

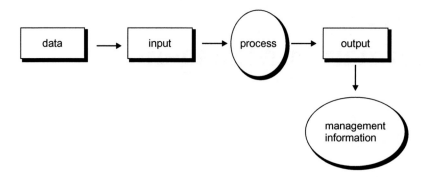

Figure 6.1 Information system in terms of input and output.

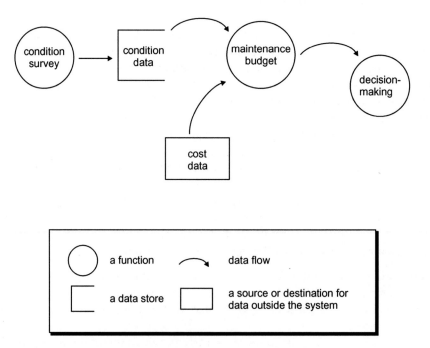

Figure 6.2 Elements of an information system.

recently can it be said to have been a major determinant. The growth of IT, amongst other factors, has raised information management to a higher position on the business agenda, so that a systems approach to the design or development of organisation structures is now more likely.

This raises an important philosophical issue as to what should be the motive force for redesigning the information system and introducing new technology. It is possible to argue that information management is the major imperative, and that the organisa-

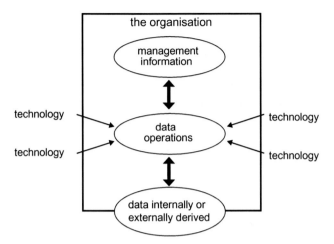

Figure 6.3 Components of the information system.

tion structure should be subjugated to its needs. Many would argue, though, that this runs the serious risk of neglecting human needs.

There is no perfect answer to this, as consideration has to be given to the characteristics of the organisation in general, and its information management needs in particular. The evolution of an information system strategy should be treated as a component of the overall strategic thinking of the organisation, rather than as an independent entity.

The development and implementation of such a strategy is a complex and delicate operation (figure 6.4), and the complexities of the process have inevitably increased with technological potential. If we think of such implementation as a project then a key issue is full and sensitive scoping of that project.

During the first generation of new technology, refinements to information management were driven by data processing requirements, and developments that replaced previous manually executed functions, such as finance and accounts. The emphasis here was on efficiency, with management information somewhat incidental. In this instance the process can be classed as a bottom-up approach.

The development of more recent technology, and the potential for an integrated approach, necessitates a much more strategic approach, and a top-down process (figure 6.5).

For the future of maintenance management this has had far-reaching implications, as the increasing tendency for maintenance management to be subsumed into facilities management leads to its being considered in a far broader strategic framework.

Information systems and technology

The development of information systems thinking is inextricably linked with the development of new technology and is perceived by professionals in terms of

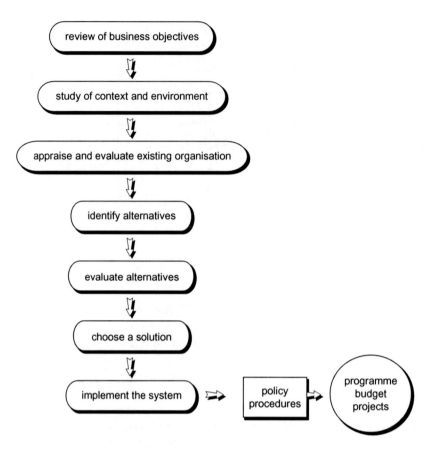

Figure 6.4 Development of an information system strategy.

PROCESS	DRIVING FORCE	EMPHASIS
bottom-up	data processing	efficiency with information secondary
top-down	senior management	information needs determined by organisational objectives
systems evolution	data users	user-orientated based on data manipulation
strategic development	top management	corporate strategy formulation with data/information integration

Figure 6.5 Implementation processes for information systems.

generations, which are paralleled within organisations by the introduction of ever more sophisticated data management applications.[1] The stages, or generations, that have been identified are outlined below.

(1) Records and filing systems were manual.
(2) Early computerisation was used to replace existing manual systems, within a very contained organisational boundary, typically the accounts department, with technology based on main frame systems and with specialised data processing departments.
(3) As applications increased there became a tendency for applications to cross boundaries, leading, in some cases, to duplication and proliferation in an unstructured manner.
(4) Development from this was a result of the search for an integrated approach, and a more formal and effective information system.
(5) Computer-based systems increasingly began to take over other key aspects, such as the monitoring and control of resources. This was accelerated by a move towards user-friendly hardware and software, and simpler data processing.
(6) Database concepts were developed that permitted data storage in a common filing system, with applications access on a controlled company-wide basis.
(7) There was a natural evolution to networked systems and mini-systems.
(8) There is now an ever increasing demand for more efficient information flow, from internal and external sources, and a move towards the automated office where systems thinking becomes a dominant force.

These stages, of course, overlap considerably and represent a quite simplified view. The value derived at each stage is perceived, in most organisations, in terms of efficiency through stages 1 to 4. Stages 5 and 6 are, essentially, seen as contributing to an improvement in performance in strategic terms. Stages 7 and 8 can be viewed as an embedding of information management within organisations, and this is likely to continue.

Data operations

This represents the essential information management aspect, and requires consideration under a number of headings.

(1) Data sources

Data can be generated internally or externally, and may be formal or informal. Formal data sources can largely be designed into the information system in general, whether internally or externally generated. Formal data is also more likely to exist in a defined data store or data source, whereas informal data may be less well organised. Much of the effort in new technology is directed towards the more efficient organisation of informal data.

Data sources can also be classified according to the point at which the data is generated, in which case it is sometimes categorised into primary, secondary or even tertiary data. This is of some importance in maintenance management.

(2)   Data collection

In its basic form, data relates to a number of events and needs transmitting to a data store. In primitive systems this would be in handwritten form and transferred to a file. In the interests of efficiency a number of pieces of paper, or pieces of data, would be dealt with together, leading to the term 'batch processing', from which derives the term 'batch input'. Input to a data store normally requires some checking process to confirm its validity, and today this procedure is likely to be electronically based, via a data-preparation phase on a computer or, increasingly, at the point of collection.

Increasingly, attempts are being made to bypass the data preparation stage by the use of optical mark reading, optical character recognition, magnetic character recognition and bar-coding linked with portable electronic devices. This in turn is likely to be linked directly to the primary information management device. Currently some research effort is being devoted to the codifying of maintenance data, to facilitate more efficient collection and effective input.

(3)   Data output

Modern technology makes possible a variety of ways of organising and processing data, and of presenting it as management information. Central to this is the need to define management information requirements. It is not too easy to generalise here, but for decision-making purposes figure 6.6 represents a hierarchical concept that is typical of many organisations.[2]

Figure 6.6 Hierarchical decision-making in the information system context.

Maintenance information needs

If maintenance is considered properly, in the context of the organisation, information needs should be defined at strategic, management and operations levels.

At the strategic level, the importance of developing an effective policy for the management of buildings requires the flow of appropriate management information in a corporate information system, especially with respect to building performance. It is a matter of some debate whether the maintenance of buildings often attracts this level of thinking.

It is at the operational level that most attention has been given to maintenance information systems and, in the majority of cases, integration with the organisational system and management decision-making has been confined to financial matters. The most common, and often the only, link was into the accounting system, and perhaps some convergence in the area of administration and office automation.

Although for many systems this still remains the case, the rapidly evolving tendency for maintenance to be treated as part of a more comprehensive facilities management operation leads to a much more sophisticated integration of the whole range of building management functions.

Accepting these complications, it is still worthwhile examining maintenance information requirements in principle, throughout the whole life cycle of a building. A simple approach is to divide the cycle into the design system and the user/occupier system. There is some merit, however, in extending the design phase to include construction, given the essential interdependence of the two. Similarly, the user/occupier system can be extended to embrace associated activity under the heading of the building management system, thus placing maintenance a little more firmly at the heart of facilities management.

Given trends in procurement, it is also the case that these divisions may no longer be particularly distinct, particularly under PFI initiatives.

The design/construction system

This phase should be considered to last from inception to commissioning and hand-over. It will possess its own internal information system (figure 6.7), through which data is collected and developed to produce what may be termed a building data store. If this is carried out properly, which represents a major information management task, it should culminate in the presentation of a package of building management information at hand-over in the form of a building manual.

However, this information system is not independent, and must link into the overall organisational information system, and access external data sources (figure 6.8). The information needs can then be considered at three levels.

(1)   User/client-sourced material that permits:
   ❏   the development of a proper brief, which is an accurate response to user requirements, and which will include an explicit statement with respect to the stance to be taken on building maintenance
   ❏   a proper decision-making process to take place, concerning the overall procurement strategy
   ❏   the physical execution of the design and construction process.

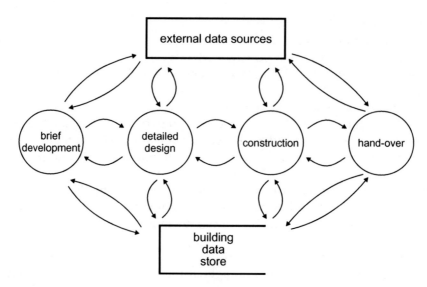

Figure 6.7 Information system during building procurement.

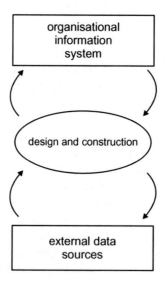

Figure 6.8 Design/construction and information systems.

(2) There are additional information needs, over and above that provided by the user/client, to permit the development of detailed aspects of the brief, design and construction. Out of these very wide requirements there will be maintenance-specific information such as:

❑   maintenance cost information and component/material performance data to allow the evaluation of design alternatives

❑ technical performance data/feedback to inform detailed design development
❑ data for incorporation into the building manual.

(3) Information required specifically during the design/construction phase. These needs can be classified in two ways. In the first place, data will be required from within the organisation, through proper feedback, which places a premium on a proper internal information system. Figure 6.9 illustrates a simple feedback loop with information flowing from a building data store into the design process. However, informal information channels may be just as important here. The link marked by a broken line could, for example, represent the involvement of a maintenance manager in the design process.

Second will be the requirement for data or information that can only be met by external sources. Although many of these are now well established, there is a major problem here of proliferation and a lack of a co-ordinated system for channelling it in the most effective way.

Taken in their entirety, these needs may be general in nature. For instance, there may be externally derived data with respect to the performance of a component. On the other hand, it may be specific to the organisation, perhaps relating to an operational characteristic or to the particular building being produced, such as a specific user requirement or activity. Specific information or data may be derived from internal and external sources, although the latter is only likely to exist with the presence of a large, well-organised user group.

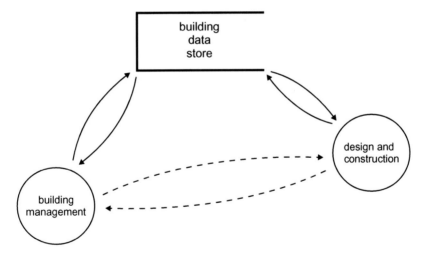

Figure 6.9 Maintenance feedback and design/construction.

The building management system

The essential link into the building management system, from the design and construction phase, is the building manual, and it is to be hoped that this will more commonly become an integral part of the organisational information system.

In terms of managing the built asset, it is necessary to take the perspectives of both the building user and the building manager. Their imperatives, and hence their information needs, will be different. The former is concerned with optimising the use of the building and the latter is charged with the execution of specific activities that will satisfy that requirement.

In optimising use of the building the user needs to know:

❑   how to use the building in the manner for which it was designed, assuming of course that this is consistent with his needs
❑   what action he takes if, in his perception, the building fails to meet these needs.

The first of these should be addressed by a properly structured building manual, to which he should have access. In practice there may be:

❑   no proper manual in existence
❑   a poorly produced manual
❑   a fundamental ignorance on the part of the user coupled, perhaps, with a reluctance to use what is available.

This highlights the general need to increase awareness of performance issues in the building user, and specifically underlines the need to produce a building manual for both professional and non-professional users.

The action to be taken in the event of a perceived failure requires a reporting mechanism and, on the face of it, this is a simple interface with the building management function. This may not always be as simple as it sounds. At the level of the simple maintenance request, the flow of information is relatively straightforward. However, underlying these requests may be a more deep-seated dissatisfaction with the overall performance of the building (figure 6.10).

The interface between the user and the building manager can be seen as a conduit through which flow:

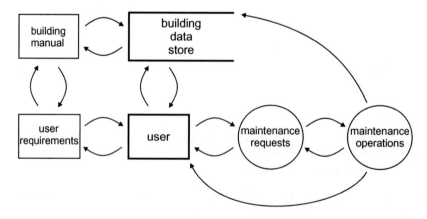

Figure 6.10 User requests and the building information system.

❑ simple maintenance requests
❑ data providing essential building performance intelligence, which needs to be
   added to the building data file and is an important component in the feedback loop
   for design teams.

The latter of these is subject to a number of problems:

❑ the absence of a proactive approach by building users
❑ failure to channel and direct data correctly
❑ shortcomings in the analysis of the data into proper management information.

There is a strong tendency for the building manager to be a passive recipient of data
rather than an active searcher for it, other than to satisfy his immediate needs. The
perfect feedback system on building performance cannot be effective without data, and
it is unreasonable to rely totally on the user to provide this.

The manager of the building requires data or information to answer four fundamental questions (figure 6.11):

❑ What needs to be done?
❑ Can it be done?
❑ How is it to be done?
❑ When is it to be done?

In determining what has to be done, the building manager may identify the following categories:

❑ requests from the user
❑ requirements derived from his own investigations, e.g. inspections and surveys
❑ regular items required by a variety of agencies, ideally taken from a building
   manual.

There are therefore three sets of data flow, all of which constitute what may be termed
the maintenance workload (figure 6.12).

In response to this the building manager must decide what can be done, which may
be limited by technical or financial considerations. These two are linked, but the latter,

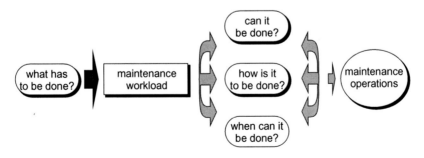

Figure 6.11 Generating maintenance tasks.

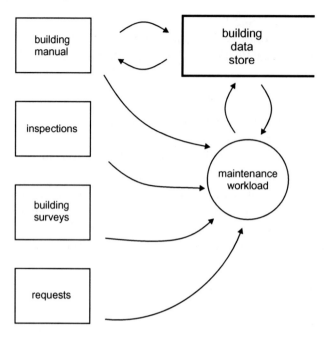

Figure 6.12 Generating the maintenance workload.

in particular, identifies the need for him to have up-to-date financial data. This requires him to be provided with a budget, and for there to be a flow of financial data permitting him to make decisions (figure 6.13). The essential requirements are to know:

❑   how much an item will cost, derived, hopefully, from a cost database section of the building data store
❑   his annual budget
❑   an up-to-date knowledge of what has been spent and committed
❑   a predicted financial out-turn.

These factors, and the question of when, are inextricably linked, and are an essential part of the maintenance planning system. Access limitations, imposed by the use of the building, frequently constrain the timing of maintenance work. The information with respect to this may be obtained from the building manual, but can also be provided by the source of the workload item. For example, planned inspections, in identifying the need for work, can also specify access limitations.

In determining how to carry out an item of work, there are both technical and operational matters to consider. The first of these requires access to technical data, which may be available internally or externally. The collection of data internally to aid technical decision-making is an important function of the information system, and it should be possible to refer to the building data store for this. There will also need to be sufficient information available so that a repair methodology can be devised, which includes

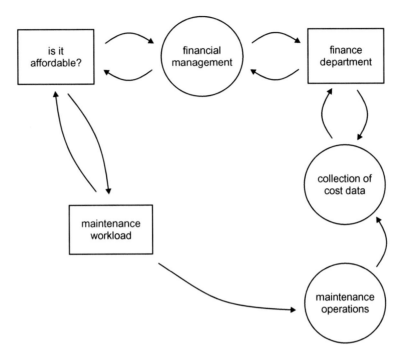

Figure 6.13 Maintenance and the financial information system.

overcoming physical access difficulties. This requires accurate up-to-date knowledge of buildings in the form of graphical information, as part of the building data store.

For operational purposes, the manager has to make a number of decisions ranging from the strategic, such as deciding whether to use direct or contract labour, down to the particular, such as deciding which contractor or maintenance operative to use. This is clearly an important part of planning, and the system set-up must provide for a steady stream of information to assist the decision-making process (figure 6.14). Additionally, the information system must provide for the proper flow of information to operative level and, following completion of the task, back into the system (figure 6.15).

Maintenance information management

Building data store

The building data store referred to is, in effect, the information systems terminology for the building performance model. Figure 6.16 summarises the data flows to and from it over the building's life. On a large estate it would be subsumed into a comprehensive model of the whole property portfolio and, as is increasingly likely, part of a more broadly based facilities management function. This store, or model, which is dynamic

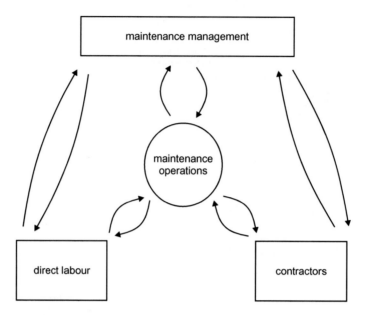

Figure 6.14 Maintenance operations information system.

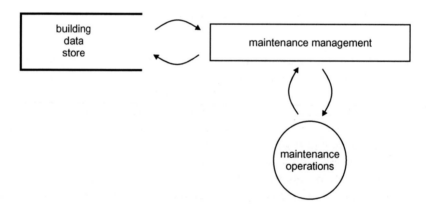

Figure 6.15 Operational feedback to building data store.

in nature, exists not only as the recipient of data generated but also as the essential internal source of data, and is part of the system for generating management information.

The close relationship of the building data store with the building manual is obvious, but they are separate entities. The manual should be seen as management information derived from the data store.

The means by which the data store is managed is a crucial issue. In some situations it may quite justifiably be a purely manual operation but, beyond the most simple of

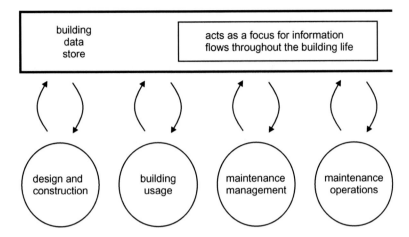

Figure 6.16 Inputs to building data store.

estates, there is a strong case, and increasing pressure for the introduction of new technology and the computerisation of the whole process. In the first instance the data-handling power of electronic data management represents a massive technical advantage and, secondly, it permits comprehensive analysis of data to produce the most appropriate management information. In addition, the process of introducing new technology imposes a disciplined approach to information management, and compels a critical appraisal of existing practices (figure 6.4).

Computer databases

Until recently, computerised maintenance management systems involved the processing of data using files, of which three generic types can be distinguished:

❑ a master file containing permanent information as a source of reference data
❑ transaction or work files relating to physical operations, and which can be used to update the master file
❑ report files, constructed by the extraction of information from the other files.

The development of ever more powerful database and operating software coupled with technological developments permits the production of fully integrated management systems. Traditionally three distinct types of database have been identified:

(1) Hierarchical databases link data in a tree-like structure. These are rather inflexible in use, but are extremely efficient for processing large amounts of data. Their use is therefore most appropriate when there is a clearly established highly structured information system in terms of input, output and data flows.
(2) Network databases are an extension of hierarchical ones but permit more cross-linking to improve flexibility.

(3) Relational databases are extremely flexible and are most appropriate for the varied requirements of managing maintenance information. In this type of database, information is organised into a series of files, but with data repeated in different files to create links between them. The very nature of maintenance information requires data to be duplicated in different groups, and a relational database permits this type of data organisation.

There are a large number of database processing applications in building maintenance. These range from off-the-peg systems to execute specific tasks, which might be termed closed systems, to loosely structured database packages, which can be customised to suit individual requirements. The choice of a system is a complex issue, involving many factors, and specialist advice may be necessary. However, there is evidence that the rather restrictive closed system approach characterising many early systems has now been replaced by much more innovative approaches that are capable of greater customisation to user needs.

The main advantages of such computerised systems are the rapid availability of information, and the ability to manipulate and organise data into a format that is appropriate to the maintenance team. It is also clear that the discipline imposed by the adoption of a computerised system encourages a systematic review of maintenance management operations, systems and strategies as long as the implementation project scoping is properly carried out. If the system is designed, implemented and embedded correctly it also provides a large measure of management continuity.

It is also clear that for these benefits to be realised management commitment is essential. This must include not only maintenance managers, but senior management in the organisation as a whole.

Integrated systems

More recent developments in computerised systems for managing maintenance have been towards fully integrated management systems, which provide direct online entry of job information, such as works orders, invoicing and payment details. This information then automatically updates property record files and contractor information. The system shown in figure 6.17 and outlined below represents a derivative model of many contemporary systems. In this sense the term 'file' is loosely used to represent a part of the data set/store forming the basis of the system.

(1) Property record file
This file contains basic information such as:

❏ tenant's name
❏ address
❏ property code.

There are then several additional sets of information that may be included in the basic file, or in a sub-file accessed from the main property record file. There is some merit in the use of sub-files as it makes for clarity, improves user-friendliness and provides a more structured approach, allowing data to be accessed at a range of levels of detail. This information may include any or all of the following information:

❑ construction details
❑ age of property
❑ details of services
❑ cyclical maintenance or inspection requirements
❑ maintenance history.

In the facilities management context there are, of course, many additional data sets that will be required – such as occupational characteristics, contents of spaces, and so forth.

(2) Contractor file

This file will keep the key information relating to currently approved contractors. There is a great deal of information that can be included and, again, it may be useful to subdivide this file into permanent and variable information. The permanent contractor information will need to include most of the following information:

❑ name
❑ address and telephone number
❑ code number
❑ type of contractor
❑ hourly rates, including out of normal hours
❑ tax certificate details
❑ insurance details.

It may also be useful to include some information with respect to the size of job that is appropriate, or the area within which the contractor operates. A typical sub-file might be a record of the jobs undertaken by this contractor, perhaps including performance data (possibly for use as part of a benchmarking exercise), permitting various analyses to be undertaken.

The example shown in figure 6.17 makes use of a works order archive file, and the data stored there may be retrieved in a number of ways, including on the basis of an individual contractor. This gives rather more flexibility, in that a code is used for the type of contractor. For example, all plumbing contract information can be extracted from this one file, allowing comparison of like with like. There are obviously a variety of ways of approaching this problem, and the one chosen will be that most appropriate to the organisation in question.

There are a number of routines that can be written to retrieve information from all these files for analysis purposes. A problem, however, with some early off-the-shelf packages is that programming basically locked access to files, other than through the routines provided with the software. Thus, flexibility was compromised. This emphasises, of course, the massive reliance on software specification at the outset and is just as true for more contemporary systems.

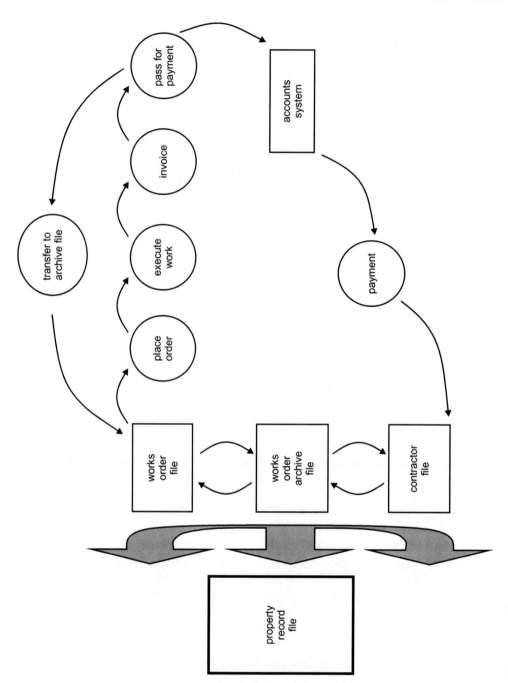

Figure 6.17 Integrated computer-based system.

(3)   Works orders

These may be raised in a variety of ways. The system described has a programme that will extract cyclical data from the property file and automatically raise an order. It is common for systems to raise orders via the computer and, in general, the following will need to be provided:

- property address or code
- contractor or contractor code
- date
- origin of order
- estimated cost
- start date and completion date
- degree of urgency
- details of work required.

(4)   Live orders file

When an order is processed, the details automatically get passed to a live orders file. There are a number of programmes that can then access live order data to provide data for management control purposes.

(5)   Invoicing

When invoice details are received from the contractor, the information is again entered into the system. The important information to be included is:

- order number cross reference
- amount of estimated and actual cost
- date.

In theory, the actual cost, date and order cross-reference should be sufficient, but processing can be simplified by duplicating information.

Once the invoice has been received and passed, the appropriate order is removed from the live orders file and the information contained in the order is transferred to update the property and contractor file. The order may then pass into an order archive file.

There may then be a number of routines for accessing archive information. The degree of sophistication possible here depends on the adoption of a rigorous operational coding system. The requirements of the information management system must always be borne in mind, as the need to use a complex coding system may be neither necessary nor realistic, given the circumstances of the organisation.

A simple example of how file cross-referencing may work is illustrated in figure 6.18.

The system described above can only be classed as an integrated information system, in that it does not fully facilitate full maintenance planning and control. Current developments are leading to the development of fully integrated management systems, and

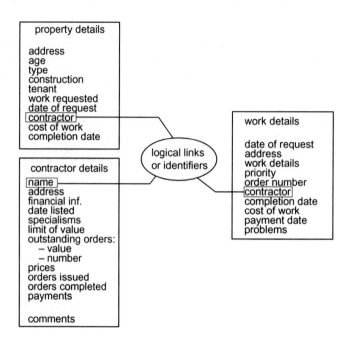

Figure 6.18 Cross-linking of data files.

at the most sophisticated level to the incorporation of CAD systems, which can be viewed as graphic databases, normally as part of more comprehensive facilities management packages.

CAD systems

Building maintenance operations require individual items of work to be carried out at prescribed locations, which are normally indicated on drawings. These may be in hard copy format, or stored electronically. Whether they can be considered to be up to date is uncertain. In many cases they may only be the original as-built drawings, provided when the building was commissioned, and in others may be so old as to require complete re-draughting following a full measured survey. If the original drawings, or the new ones, are CAD based then their regular updating is greatly facilitated.

Alongside this graphical information will be a large body of text data, either in a building manual, on a computer database or in an integrated system such as that described above.

Consistency between graphic and text data is often very questionable, and there has more recently been a move towards a more integrated approach, particularly in the facilities management area.[3] By attaching database information about the building, or facility, to the graphic entities in the drawing, the database can be accessed via the on-screen drawing.

AutoCad™ is a particularly widely used CAD system, and many peripheral software packages have been designed to integrate with it. The system acts as an electronic drawing board, and can be used in two- or three-dimensional mode. Current versions allow the direct introduction of building components such as walls, roofs, windows and doors, selected from a library of standard components. In addition, many component manufacturers now produce product information on a disk, for incorporation into AutoCad™.

For a CAD system to be used effectively, the information contained in its files must be organised in a systematic manner. This is to ensure the efficient transfer of information between different members of the design, construction and management team. This, therefore, requires the adoption of a standard method of data organisation.

Part 5 of BS 1192 gives guidance on the structuring of computer graphic information, with the aim of simplifying its transfer. It identifies two methods of data organisation in CAD systems:

❑   the use of sub-models
❑   the use of a layering system.

The use of the former requires the adoption of a hierarchical approach, which is less flexible than layering systems, which are now almost universally employed.

Sets of graphical items that make up a component, or part of one, are termed 'entities'. When these are entered into a model, they are assigned to a 'layer'. This can be visualised as a transparent sheet of paper. The full CAD model thus exists as a series of layers, or transparencies. Each layer can be activated or suppressed by the operator, on the screen, or in a plot.

Information can thus be grouped logically on an appropriate layer, and a high level of selectivity is available. As well as reducing clutter on drawings, this provides a very powerful information management tool. It is quite possible to name layers in any way, but standard naming conventions permit easier transfer of data between applications and users.

An integrated computer model

Software packages such as AutoCad™ are, in themselves, very powerful data managers, and have the ability to present information in other than graphical form. For example, there are schedule generation facilities, which provide for the extraction of attribute data into an organised form. This data can also be extracted for processing by a range of external data management packages.

A number of facilities management groups have begun to exploit the potential that exists, and develop fully integrated systems. The key characteristics of typical contemporary systems are described below.

A major obstacle to progress for some time was a failure by most design teams to use CAD in the design and commissioning of buildings. It is at the design stage that the formative model is best constructed, although, as we noted earlier, even if this exists its appropriateness in many cases for building management is still questionable and there are also continuing issues with respect to shelf life. Keeping such models up to date requires a systematic approach to information management, but if this is neglected they become very expensive to update.

Contemporary systems

There are a number of packages currently available at a range of prices that offer integrated facilities management systems. Typically these will include provision for:

❑   space occupancy management and auditing
❑   asset inventory operations
❑   personnel management features
❑   service centre/helpdesk facilities
❑   condition management including planned maintenance and automated data collection links
❑   works order management and invoicing
❑   overall financial management and budgeting arrangements
❑   customer satisfaction surveys and analysis
❑   the ability to produce benchmarking data
❑   use of a CAD platform

Archibus[4] produce a web-based Integrated Works Management System, ARCHIBUS/FM Web central, that provides a highly sophisticated data management system for managing a portfolio of buildings, making use of the latest communication devices.

Whilst most systems make use of a CAD platform, Decision Graphics[5] in their new AutoFM Desktop abandoned the need for a CAD platform on the basis that experience showed that 80% of facilities management applications comprise database information and only 20% graphics information. The system can still relate data into CAD but provides a Windows 95 and NT-based system that can be used independently.

The majority of systems, however, integrate database software and a CAD facility. Fast Track[6] PPM from Intelligent Information Systems (iiS) is a relatively low-cost product, but one that is adaptable in terms of scale of use, dedicated to maintenance management although its origins are in the facilities management field. It can thus be thought of as an integrated system. It provides for:

❑   scheduling of maintenance, service and inspection tasks
❑   centrally managed issuing of works orders
❑   invoicing
❑   a comprehensive range of reporting functions.

Help desks are a key feature of the majority of contemporary systems and Fast Track provides a service desk facility that:

❑   logs and records calls centrally
❑   manages the logged calls and issues works orders

❏ makes use of common tasks or predictable information through drop-down lists
❏ integrates requirements of service level agreements.

Communication in this system can be simply through e-mails with attachments for equipment lists, maintenance manual references, and Health and Safety guidelines, for example.

As an additional module to this package, Help Desk and More provides a condition survey and inspection management facility using hand-held digital equipment for data collection, which is uploaded into the main system which then processes and analyses the data and schedules work.

Fast Track is essentially a modular approach that provides the maximum flexibility. Individual modules can be run on their own, on existing networks or as a web-based solution using iiS's Facilities Net.

An interesting approach is taken by FMx Ltd in their CAFM Explorer software.[7] There are a number of key features of this highly sophisticated product that have been developed over a number of years with the assistance of a very active user group. It is designed as an open system through a software development kit that permits the user to adapt the standard product to their needs. CAFM Explorer is essentially a powerful database that stores, retrieves and distributes data through the whole range of electronic media. Although FMx Ltd acknowledges the predominance of non-graphic data, its nature permits it to embrace AutoCAD where appropriate. Unlike other systems, it is not modular and its flexibility comes from its open nature and programmable interface so that it can readily connect with other parts of the business organisation, such as finance or personnel.

A key feature of the software is the user-friendliness provided by an interface that has a Windows Explorer appearance. This provides for very simple navigation around the system.

Information sources

Sources of information for maintenance work can be divided into internally and externally generated material. Maintenance data, and information, can also be separated into that which is of a general nature, and that which is specific to a particular building or task. The former is most likely to emanate from an external source, whereas the latter will probably be internally generated. Large estate owners, however, may generate data having more general application, either internally or externally. The maintenance of good external data sources also requires a flow of data from individual organisations to produce a sensible data bank.

There are three major issues to be considered.

(1) Information channels may be formal or informal, and the importance of the latter must not be underestimated. The major problem, however, is one of harnessing and organising this material.

(2)   At a national level there is a proliferation of material, and this presents a major organisational and analysis task. Transfer of the information to the most appropriate point of use is a well-known problem.

(3)   Raw data, whichever direction it flows in, internal to external or vice versa, requires manipulation into meaningful information, and this has to be executed with a great deal of caution. Information of a general nature often only represents an average stance.

This means that, on a national basis, the industry is confronted with a critical information management task.

Internally generated information

The flow and storage of data throughout the life of a building, from inception into occupation, has already been discussed. The main vehicle for collection of data is the building data store, or building performance model. Such a facility will always exist, in a form ranging from the primitive to the technologically sophisticated.

The specific items of data and information generated in this way are illustrated in figure 6.19, where the categorisation used is by point of generation. This is essentially a continuous process, and the illustration represents a simplified model.

Information from external sources

The following external data sources are of significance.

(1)   Specific government departments have responsibility for managing the buildings that house their various functions, and these provide data and information on the performance of their buildings in use. They will, for example, give design guides and recommendations on maintenance procedures.

In addition, there is a limited amount of cost-in-use data available, and they may transfer some of this data to other collecting agencies.

(2)   Local authorities generate information on many buildings for which they are responsible. Although little of this information is published, it finds its way to other agencies. A major contribution is with respect to housing maintenance, and there are a number of formal and informal groupings that provide valuable collection and dissemination points.

(3)   A significant portion of what may be termed 'social housing' now rests in the hands of housing associations, who are overseen by the Housing Corporation. The Housing Corporation, through the financial control it exerts and its role as an advice centre, provides important data.

(4)   The National Audit Office is a government watchdog, charged with appraising the effectiveness of a range of public sector organisations, with the Audit Commission having specific responsibility for Local Authorities in England and Wales. Whilst their reports tend to represent overviews, they are of great value in terms of the principles they outline and the guidelines they lay down.

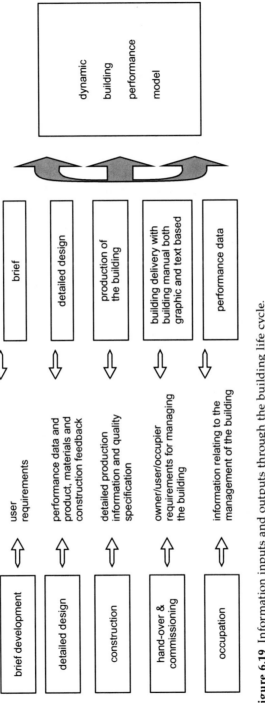

Figure 6.19 Information inputs and outputs through the building life cycle.

(5)   The Building Research Establishment is an important source of information for all sectors of the industry.

(6)   The Chartered Institute of Public Finance and Accountancy publishes annual statistics for the maintenance costs of public housing, and advises on accounting procedures that may be used for this type of work.

(7)   The British Standards Institute publications, whilst of general applicability, provide important maintenance-specific recommendations. The time taken in drafting standards means that they do, at times, tend to lag behind current practice.

(8)   The British Board of Agrément was set up in 1966 to provide technical assessment of new products. Their certificates of approval include an assessment of the probable performance of the product in its intended use and, in appropriate cases, the maintenance requirements and probable life. Although of particular use for maintenance purposes, they only cover a limited range of products.

(9)   Manufacturers' information is produced in large quantities, although its quality, in information terms, is extremely variable.

(10) Trade associations, set up to promote the proper use of specific types of materials and products, provide useful maintenance information, and will often give advice on specific problems.

(11) Professional bodies have maintenance interest groups, which produce a range of valuable publications. One of the most important of these is Building Maintenance Information (BMI), an RICS company.

(12) There are a number of other bodies that either directly or indirectly support maintenance activity. These include the Joint Contracts Tribunal, the Facilities Management Association and the British Institute of Facilities Management.

Most of this external information will tend to be of general or background significance, and its interpretation may be difficult. A major issue is the means of accessing this information, which is now increasingly becoming automated. It also raises the question of a suitable means of classifying information.

Classification and coding of maintenance information

General information

Most general information will be generated externally, and will be classified using industry standard methods. The two most relevant classification systems are:

❑   Universal Decimal System
❑   Uniclass (Unified Classification for the Construction Industry).

The former is rather rigidly hierarchical and, whilst useful for accessing published data, is of little use as an aid to data manipulation.

Uniclass was developed to replace the CI/SfB system and provide the construction industry with a classification scheme for organising library materials and structuring product literature and project information. Whilst the initial focus has been on information management relating to new build, it should become an important tool for information systems relating to maintenance activities.

Also of relevance here is the Code for Production Information produced by the Construction Project Information Committee.[8]

Specific maintenance information

Very specific maintenance information does not lend itself to classification using either of these methods alone. However, internally generated reports on the performance of a particular product can easily be given a Uniclass classification. The major classification problems, and hence coding difficulties, stem from the characteristics of maintenance operations.

Not only is maintenance information extremely fluid, in the sense that each stage constantly updates and passes on information, but there are a number of characteristic items of information associated with a maintenance operation:

❑   location of the operation
❑   building element or group of elements
❑   function of element/location
❑   reason for the operation
❑   description of the operation
❑   who did it and when
❑   magnitude or seriousness
❑   frequency of occurrence
❑   resources expended and time taken
❑   budget and actual cost
❑   source of funding.

There are any number of combinations of these that can be used to order and group information, all of which require a coding system of one sort or another. In a sense, these items represent the raw data of maintenance activity, and their grouping, sorting and coding is a fundamental information management issue. This means that the degree of coding sophistication should be determined by properly defined management information requirements. If independent management functions are considered in isolation, a simple approach will suffice.

For everyday operational management, sort criteria, based simply on location and work gang, may provide an appropriate basis. Alternatively, a combination of location and element can be used as a basis (figure 6.20). This approach has some merit also, in that it is consistent with the way in which inspections and condition surveys might be carried out, which opens up the opportunity for an integrated electronic data capture and storage approach.

The structure outlined in figure 6.20 can be supplemented by a code that describes the function of a space. On the other hand, for financial management purposes it may be necessary to group items in terms of cost centres, or accounting periods.

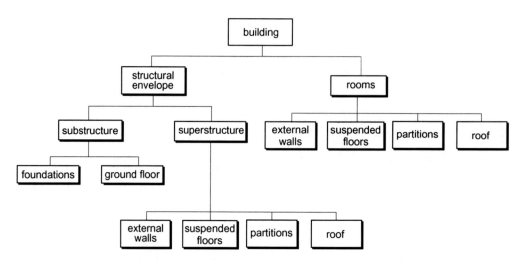

Figure 6.20 Information classification based on location and fabric element.

For design feedback purposes, elemental or functional criteria may be the most appropriate. There are, for example, a number of published information sources in the form of elemental cost analyses of buildings in use. Four structural approaches have been proposed:

❏   those paralleling the ones used for new-build purposes
❏   those designed to facilitate the collection of data
❏   those that are an extension of accounting procedures
❏   those designed to permit cost comparisons of maintenance expenditure incurred by different organisations.

In designing an integrated management information system, it is apparent that there may be a number of conflicts. This highlights the usefulness of relational databases, particularly when management requirements dictate the necessity for an analytical capability on a large estate.

This requires the use of a coding regime at the data source – that is, the execution of the work. The coding of items themselves, however, has to emanate from management. The data collection demands at the workface should always be strictly limited. The prime requirement at this point will probably be the collection of labour data, and perhaps material usage. For a number of reasons, this data needs to be treated with caution.

The most appropriate methodology, at the present state of the art, appears to be a sort code approach. An example is shown in figure 6.21. This is not incompatible with the layer-naming strategy associated with CAD systems. Obviously, however, such an approach becomes increasingly cumbersome with increasing numbers of fields.

Other approaches involve the development of a standard maintenance operation phraseology, which may have benefits in terms of describing actual operations. To some

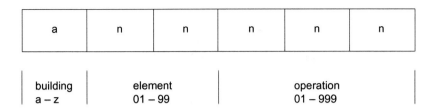

Figure 6.21 Use of sort code for coding maintenance information.

extent, a standard phraseology does exist in standard schedules of rates and specification clauses. Many computerised maintenance management systems will incorporate, or link with, such documents, particularly for costing and budgeting purposes.

References

(1)   Edwards, C. *et al* (1991) *The Essence of Information Systems*. Prentice-Hall, London.
(2)   Edwards, C. *et al* (1991) *The Essence of Information Systems*. Prentice-Hall, London.
(3)   Lovejoy, D. (1990) Getting the most out of CAD, *Facilities*, 18(12).
(4)   www.archibus.com
(5)   www.aecnews.com/articles
(6)   www.iiSFM.com
(7)   www.cafmExplorer.com
(8)   CPIC (2003) *Product Information: a code of procedures for the construction industry*. RIBA, London.

Chapter 7
Maintenance Planning

Introduction to planning principles

The process of planning for maintenance work has much in common with the planning of any construction activity. Therefore, the basic principles of planning should be firmly understood before considering maintenance planning specifically.

Essentially planning must be seen as a thought process. Whatever activity is engaged in, whether consciously or subconsciously, some plan is formulated mentally. In many cases there will be no formal commitment on paper, but an intellectual process will have been engaged in to get from point A to point B, or to make product X.

As the nature of the product or activity becomes more complex, a point is reached where it becomes necessary to commit some, or all, of this plan to paper, and a formal programme is produced. At a simple level this may only involve writing dates into a diary, whilst at a more advanced level the use of a powerful computer-based management technique may be necessary.

The point at which the transformation from a simple representation to a more sophisticated one occurs is imprecise, and dependent on a great number of factors, not all of which are necessarily related to the complexity of the task being planned. The use of sophisticated planning techniques may appear as something rather clever, but in reality they are only as good as the thought processes underlying them.

Objectives of planning

Planning, as an intellectual process, permeates all activities in one form or another, always with some objective in mind, whether or not this is overtly stated. The clear identification of objectives is an essential pre-requisite of the whole process, but particularly prior to the committal of a plan to the formal programming process.

In the construction industry, planning has all too often been afforded insufficient credibility. In many cases this is because not enough attention is given to the purposes for which a plan is required, leading to a failure to produce programmes that are consistent with the planning objectives. This tends either to the bringing of the planning process into disrepute, or the setting up of an intensely bureaucratic management regime.

Essentially, a planning team is concerned with the provision of management information, and the purposes for which this information is required will be the overriding factor in determining the most appropriate means of producing it. A systematic review of needs leads to the development of a number of key descriptors, which summarise the essential requirements of a planning system.

A major function of project management is that of control. To control something implies measurement, comparison with a benchmark, drawing conclusions and taking appropriate action (figure 7.1). An objective of project planning, therefore, may be to lay down a formal benchmark in the form of a programme, to act as part of a control system.

A number of aspects in a project may require control. The most obvious of these is time, and it is with this that construction programmes are normally concerned. However, these programmes are not exclusively time-controlling tools, and they may need to be used for a number of other purposes. There will, for example, be an interest in controlling the rate at which money is spent, so cash flow forecasts and budgets will need to be produced. This is, of course, of major importance in maintenance management. In order to produce a financial plan a temporal plan is required, so these two are intrinsically linked.

There are other aspects of a project that need to be controlled, such as quality. Quality benchmarks are set through drawings and specifications, and these can be viewed as representing a model of the project. In the same way a programme can be viewed as a temporal model of a project.

In order to be successful, a programme needs to provide as accurate a predictive temporal model of the project as possible, against which real progress can be measured and assessed, so that the *status* of the project can be determined. Within a proper control system the programme must be able to extrapolate forward from measured progress and thus be *predictive*.

The process is essentially an ongoing one throughout the project, and as a control system implies the taking of action, it is useful if the planning system also acts as a *diagnostic* tool: for example, how the current status was achieved – that is, the programme provides a *historical record* that can be used to carry out a diagnosis of project difficulties.

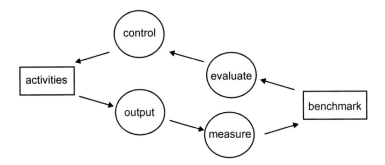

Figure 7.1 Simple control system.

If the programme is to be successful in providing a temporal control model, it must be *appropriate* for management purposes, and should include information that is both *relevant* and *realistic*.

The planning process does not end with the commencement of a project, so a good system should be *dynamic*, in the sense that it moves with the project. Too often construction programmes are produced under a very rigid regime with a quite unreasonable reluctance to revise them as circumstances change. If the programme is a genuinely dynamic model, it has to be capable of responding to changing events and be *flexible*.

As a management information service, the planning system has to act as an aid to decision-making and permit management to ask questions and evaluate the consequences of alternative actions. In order to provide this service, planning systems should be *interrogative* and *interactive*.

Overlying all of these requirements is the need for a programme to present its information in a clear and concise way, and thus be *communicative*.

The achievement of these requirements, represented by the italicised keywords, depends not only on the skill of the planning team, but also on choosing the most appropriate planning techniques, and for this reason it is important that the components of a programme are properly understood.

The components of a programme

In the discussions below the term 'project' is used, and this should be viewed in a very broad context that includes the planning of maintenance work. Four distinct but inter-related components of any plan can be identified.

(1)   A number of discrete activities or tasks
Whether viewed as a thought process, or as a formal plan, any project will comprise of a series of steps that have to be taken in order to move towards the attainment of the end goal. In a programme these will be represented by a series of what are normally referred to as 'activities'. The preparation of a programme, therefore, requires the breaking down of a project, which is in reality a continuum, into discrete activities. This is often considered a fairly simple, if laborious, exercise to be dealt with as easily and quickly as possible in order to move on to the 'real' task of playing with logic diagrams and computer displays. In actual fact it is at this initial stage that the real skill of the planner is tested most severely, as there are a number of questions that have to be resolved.

In the first instance, projects do not necessarily divide themselves into discrete self-contained activities, so that divisions need to be considered carefully. Too often insufficient thought is given to this process, and decisions are made in a somewhat arbitrary way. On the average construction project, for a new building, a series of activities can be readily identified, and the assumption is then made that these can be linked together into an orderly sequence that represents the construction process. In reality this is a simplistic and almost completely erroneous view, as experience shows that activities overlap, are discontinuous and more often than not do not proceed in an orderly sequence. This leads to a quite unrealistic assumption about the nature of the planning process, and it is essential that a planner is aware of the limitations of what he produces.

The way in which the project is broken down is, therefore, of great importance to the success or otherwise of the planning process. Initially, an answer is required to the fundamental question as to how many activities are necessary, and this will determine the level of detail of the programme produced.

To answer this question two important issues have to be addressed:

❑   the purpose for which the programme is being produced and the management objectives attached to it
❑   the quality of data, in terms of detailed knowledge of the project, available at the time when the programme is being produced.

Programmes are produced for many purposes, and it is unlikely that one detailed programme will serve every purpose. For example, a programme may have insufficient detail because the programme is not broken down into enough activities, whilst in other cases a fully detailed programme may be inappropriate for a senior manager who only requires an overview.

With respect to data quality, the uncertain nature of construction activity has to be recognised, and it must be accepted that, in the early stage of a construction project, only a limited amount of detailed information about a building is likely to be available. At this point any attempt to plan a project in minute detail, as well as being inappropriate, will be futile because it will inevitably involve making assumptions that will be overtaken by events, sometimes before the programme has been circulated.

The level of detail of the programme must therefore be consistent with the quality of information available at the time of its production and appropriate to the objectives for which it is being produced.

As an illustrative example, figures 7.2, 7.3, and 7.4 show, in simple bar-chart format, three programmes that might be produced by a contractor for the same project, at various stages in its life, for different purposes.

The major challenge associated with planning maintenance work is the need to deal with a large number of often small activities of a jobbing nature. This will require maintenance to be considered at a number of different levels, ranging from the operational to the strategic, where the level of detail in terms of management information will vary considerably.

(2)   Timescales/activity duration
Having broken down the project into a series of activities, the next requirement is to place a timescale against each of them. In the construction of new work there are several schools of thought.

❑   Information from bills of quantities may be used to work out activity duration, but there are numerous reasons why this is not a good practice.

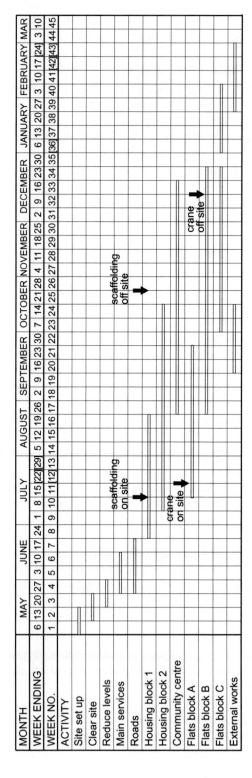

Figure 7.2 Tender programme.

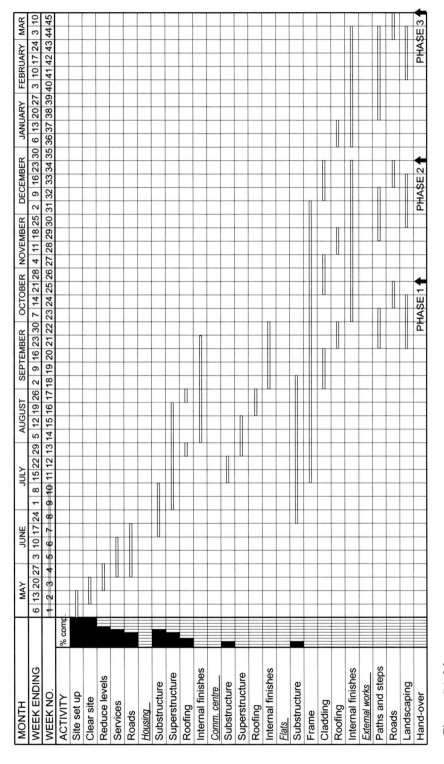

Figure 7.3 Master programme.

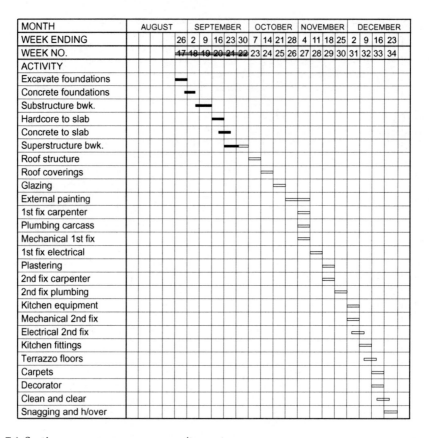

Figure 7.4 Section programme – community centre.

❑ Work study data was traditionally an option but has becoming increasingly infrequent for new construction work. Its principles may still have some currency for prediction of timings and costing for repetitive items of maintenance work where some sensible data may be available.
❑ A mixture of assessment and experience may be used which, given a healthy degree of realism, may be as good a method as any other.

The nature of new-build work, from the contracting point of view, has changed so rapidly that it is now virtually the norm for the main contractor to subcontract the whole job in various packages, so that project planning and control is less to do with in-house productivity and more to do with co-ordinating a group of separate workforces. Within this context the determination of activity durations tends to become somewhat politically influenced.

Maintenance work clearly differs somewhat from new work, and to some extent a good planning system may generate its own data in order to enable sensible predictions about activity duration to be made. A much more difficult problem in respect of maintenance work concerns not the individual task duration, but the prediction of appropriate cycles for regular activities.

(3)   Sequence of activities

A large part of the mental process of planning is related to the ordering and sequencing of activities in a logical manner. The problem here is that the translation of a mental process, which might well view a project as a continuum, into a formal programme statement tends to require the breaking down of the project into a series of discrete activities. This is fraught with the difficulties identified above, and in order for a plan to be truly dynamic, enabling it to be used for diagnostic purposes, these discrete activities should be formally linked in some way.

If the programme is to operate properly, within an effective control system, then an accurate picture of the current state of a project is essential. To derive this it is necessary to know the effect on a project of the state of progress of a number of activities. In practical terms this presents something of a dilemma.

A range of planning techniques are used for building work, ranging from the simple bar chart to sophisticated network systems. The former has the advantage of excellent visual presentation but, in its simple form, activities are not linked, so that it is difficult to assess the overall position of the project. This is demonstrated in figure 7.5.

On the other hand, a network, which satisfies this requirement, is, in its purest form, a poor visual communicator. The cheaper and faster computer hardware and software programmes, now available for construction work, overcome this deficiency by having clear graphical representations, thus permitting the potential benefits of networks to be exploited.

There is no doubt that the use of networks does have many benefits, not least of which is the discipline they impose on the thought process of planning. Most maintenance work does not, however, justify their use, although there is a need, in some cases, to consider the interrelationship of activities.

In general the overriding imperative in maintenance planning is to communicate information clearly, and this will determine the choice of methodology.

(4)   Recording of progress

The final ingredient in a plan or programme is a method of recording actual progress against that which was planned. Two ways of achieving this are illustrated in figures 7.3 and 7.4.

Of great significance in respect of recording progress and improving presentation has been the development of user-friendly, readily available computer software that provides the means to carry out analytical exercises with much more freedom. The major issue, therefore, given the characteristics of maintenance work, is that of data management, and this is a question not only of choosing the correct database or software package, but also of structuring data in the most appropriate fashion.

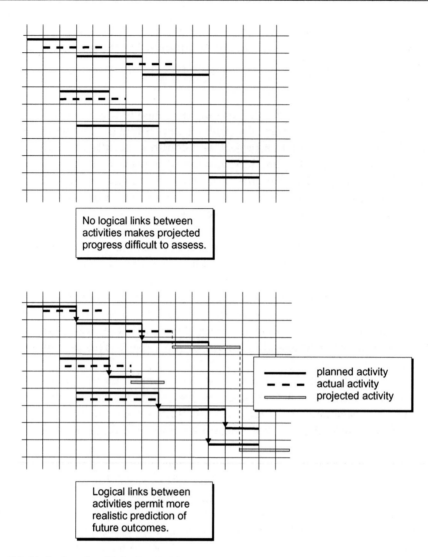

No logical links between activities makes projected progress difficult to assess.

planned activity
actual activity
projected activity

Logical links between activities permit more realistic prediction of future outcomes.

Figure 7.5 Linked and unlinked activities.

Maintenance planning

Scope of maintenance planning

Previous chapters have identified operations that may contribute to a maintenance workload and introduced the concept of planned maintenance. However, there are a number of aspects of maintenance that require planning, which may not necessarily be part of a formal planned maintenance programme. For example, it may have been decided to institute a programme of planned inspections to verify that statutory requirements are being fulfilled, or considered prudent to operate a planned replacement policy, as part of a preventive maintenance programme. This may operate separately from an ongoing planned maintenance programme.

Both these will require the maintenance management team to engage in an analytical planning process to determine appropriate cycles for inspections and replacements, concurrently with suitable presentation methods. The process requires a scientific approach, demands good data, and demonstrates that the real scope of maintenance planning is much wider than simply plotting a series of cyclic activities on to a bar chart.

Planned versus unplanned maintenance

Within any maintenance organisation there will be planned and unplanned work. The balance between the two will vary, depending on the nature of the organisation and its attitude to building maintenance. There will be an optimum balance between them that is right for the organisation concerned. A low level of planned maintenance in an organisation does not necessarily reflect a poor attitude, as it may be appropriate for the given situation.

It is quite possible to envisage a scenario where the introduction of a sophisticated planned system is not justifiable. For example, the owner of an estate consisting of one relatively simple building may choose to carry out all maintenance on demand, and plan only relatively obvious items, such as a redecoration every four years. Even the latter may be on a rather ad hoc basis. This closely mirrors the approach of the owner/occupier of a dwelling house, and is an inevitable consequence of work that is characterised by a large number of relatively small, low-level operations and a small number of larger ones. The latter are more likely to be foreseeable ones, and hence planned for. They are likely to fall into two categories.

❏ A regular ongoing requirement to perform certain operations, such as redecoration. These tasks will tend to be cyclical in nature and, in theory at least, quite conveniently form part of a rolling programme.
❏ Major renewal or repair projects which from time to time become necessary. For example, there may be a programme instituted by a housing association to replace all flat roof coverings over a fixed time period.

Some of these larger exercises fall into the category of what may be termed preventive maintenance, and need to have been subjected to a rigorous decision-making process. For example, a decision to replace flat roof coverings ahead of failure is a preventive measure. In reaching this decision, account would have been taken of the disruption and possible consequential damage of not replacing until failure had occurred.

A large number of very small jobs cause complications, in that it is not easy to make a case for a planned replacement or repair as part of a preventive policy for every item that may go wrong or wear out. An example of this might be the replacement of ball valves.

There may well be circumstances when a planned replacement policy is viable, but in normal work this is unlikely. Thus, a large number of items might be considered as unpredictable and difficult to plan, and in a very small estate this may prove to be quite satisfactory, provided the response time to a maintenance request is short. However, in a large estate it can result in both a poor service and an inefficient one. A policy of planned inspections of property, on a regular basis, provides the means by which a proportion of otherwise unpredictable items may reasonably be anticipated and placed into a planned programme.

The advantages in a large estate of maximising planned maintenance can best be visualised by considering what happens if little or no planning takes place. The maintenance workload will be characterised by a large number of small jobs, mainly of low value, distributed over a wide area and occurring in an unpredictable pattern. Not only will these jobs be disproportionately expensive to execute, but there may also be an unacceptably long response time. To provide a highly responsive service, in such circumstances, there would have to be a pool of labour ready to respond on demand.

This is clearly expensive, as illustrated diagrammatically in figure 7.6. The consequences would be a high proportion of non-productive time, probably high administration costs, and a work programme that is extremely difficult to control.

Optimising planned maintenance requires a substantial amount of decision-making to take place with respect to work cycles and replacement. If executed properly, planned maintenance work will be carried out in a more efficient way, as items can be grouped

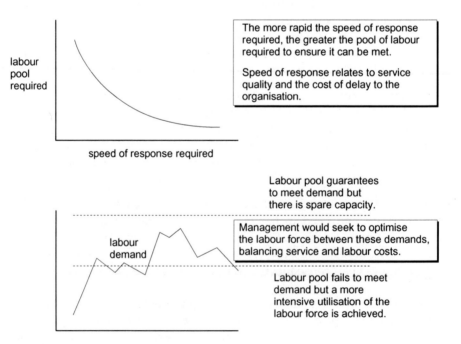

Figure 7.6 Effect of response times on labour pool.

and the advantages of economies of scale can be realised. However, even with a rigorous analytical regime and precise inspection policy, there will be an element of unpredictability. The problem that then occurs is that the incidence of unpredicted demand may take resources away from the planned programme. This will have two consequences. In the first instance, a decision will have to be made with respect to the speed of response that is required, and the implications this may have on resources. Secondly, there will be a need to decide which planned item is to be taken out of a programme to resource the unplanned one.

This situation emphasises the need for the inclusion, in the maintenance plan, of some form of priority coding, coupled with provision of a sensible contingency to allow for non-planned work.

Good maintenance records and historical cost data may help provide the latter, so that the maintenance plan strikes a balance between these interacting but conflicting workloads, and consists of planned or scheduled work alongside a contingency system (figure 7.7).

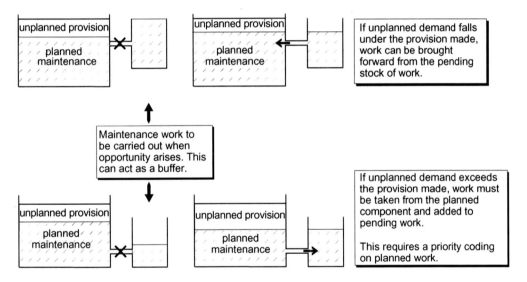

Figure 7.7 Planned work and contingency for unplanned work.

Planned inspection cycles

Planned inspections may be driven by a variety of factors, including statutory requirements. The frequency of some of these inspections may be clearly prescribed, particularly in the case of mechanical plant. These cycles may, however, be much shorter than would be dictated by fabric considerations alone and, to complicate matters even further, they may themselves all be different. For example, some items of machinery may require six-monthly inspections and others annual ones. A conscious decision may be made to inspect everything at six-monthly intervals, as the additional cost may be minimal in terms of the potential to identify problems before they occur. The appropriate frequency of inspection of the fabric of the building is by no means easy to optimise, and may or may not coincide with the foregoing. It is likely, therefore, that there will be a number of different inspections required at a variety of intervals.

Once the periods between inspections have been identified, it should be possible to co-ordinate them. However, the information collected during an inspection may suggest revised future inspection cycles. For example, it may be noted that an item is deteriorating to the extent that its replacement is becoming likely, but with some uncertainty attached to the diagnosis. This might suggest a future inspection on a shorter cycle than had hitherto been the norm.

Practical considerations will also often dominate the timing of an inspection, such as is the case when an inspection is carried out as part of a physical maintenance operation.

It is possible, if appropriate data exists, to build a mathematical model to determine an optimum inspection cycle for one item. To carry this out for every item included in an inspection routine would clearly be a major exercise, and not normally justifiable. It is, however, instructive to consider how such a model might be constructed, as it demonstrates the factors that need to be considered.

There is a presumption that a regular inspection policy will reduce emergency repair work. The first benefit that accrues, therefore, is to reduce the cost of a repair or replacement by executing it within a planned programme, rather than as an emergency item. Further benefits proceed from this, in that there will be a saving in costs associated with disruption to building usage caused by element failures. However, on the negative side the inspection has a cost attached to it, and may itself impose disruption. A statistical model, based on gully inspection, is shown in figure 7.8. It takes into account the following variables:

❏   the average number of call-outs per annum
❏   the frequency of planned inspections
❏   the average cost of an emergency repair
❏   the average cost of planned inspection
❏   the number of gullies in the inspection programme.

Another variable that might be included is the consequential cost of a gully becoming blocked in the event of an unpredicted failure.

It is necessary to make an assumption about the relationship between the number of emergency callouts and the frequency of inspection. In general in this sort of situation, an exponential relationship, as indicated in figure 7.8, is assumed: that is, the number of call-outs reduces as the frequency of inspection increases, but not linearly. To carry out a full analysis, a considerable amount of statistical data is required for the variables identified above, and the actual equation of the curve shown in figure 7.8 has to be determined.

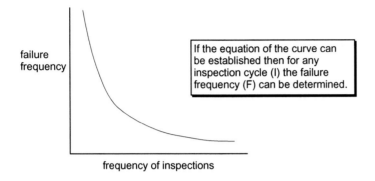

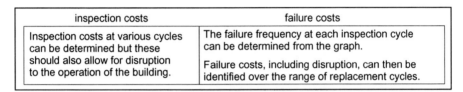

Figure 7.8 A model for optimising inspecting cycles.

Replacement decisions

Replacement decisions tend to divide into two categories, which are not mutually exclusive. In the first instance, there are replacement decisions based on the known cost of continuing to repair an item. The problem is to determine when the time for replacement has come or, better still, to predict when it will occur. There are many instances when this decision will be a professional one, based on feedback from a programme of planned inspection. For simple items, the decision may be an easy one, requiring the inspector to judge against a simple criterion.

It is also possible to take an analytical approach, and this may be justified in the case of more costly items. Provided cost data is available on historical and future maintenance costs, and the cost of replacement, a discounting exercise will help decision-making. Probably the best approach in such circumstances is to convert the future replacement cost to an annual equivalent revenue cost which, when combined with predicted maintenance costs of the new item, can be compared with the cost of maintenance if nothing is done. This technique can be adapted to make a decision now on whether to replace, or to help determine an optimum replacement date in the future.

As noted earlier, discounting techniques need to be treated with caution and depend on the availability of adequate data. Even if maintenance records are sufficiently well developed to provide basic repair costs, there are other costs and/or benefits that need to be considered. For example, a repair or replace decision needs to take into account the possibility of a complete failure that will impose disruption costs. These will probably be even more difficult to predict with accuracy.

The other approach to replacement decisions is to adopt a policy of planned replacement on a cyclical basis. This can also be modelled statistically to assist managers. The example shown in figure 7.9 is intended to assist in deciding whether or not to replace light bulbs on a cyclical basis, rather than on demand.

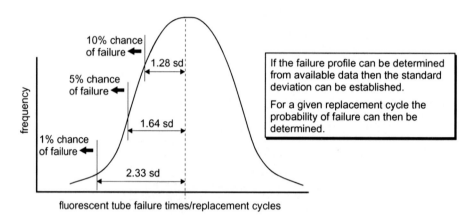

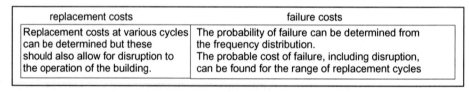

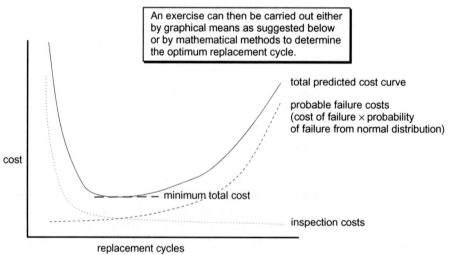

Figure 7.9 Replacement cycle model.

This type of technique requires data on:

- [ ] the failure profile of bulbs
- [ ] the cost of individual re-lamping, which needs to include travelling time and the cost of accessing the bulbs
- [ ] the cost of bulk replacement, which should enable benefits of economies of scale to be derived
- [ ] consequential costs of an individual bulb failing.

It is not difficult to envisage situations where a planned replacement policy might pay off. For example, in an industrial building with very high ceilings and ceiling-mounted lighting fittings, access for replacement may require the use of a tower scaffold, so that the cost of the failure of one fitting can be close to the cost of a complete replacement programme.

To make such a policy feasible the decision-making process will also have to determine the optimum replacement cycle. This will be very heavily dependent on having an accurate failure profile and the consequential cost of failure.

Appendix 1, in figure A1.19, and figure 7.9 illustrate how this may be approached, and show how management has to make a decision as to what represents an acceptable probability of failure, taking into account an appropriate balance between the cost of failure and the cost of a planned replacement.

Planned maintenance cycles

There will be a number of items whose inclusion in a planned maintenance programme is unquestionable and, for these, the main requirement will be the determination of an appropriate cycle period. The obvious example is decorating frequency. There is a considerable amount of data available on decorating cycles,[1,2] and a number of decision models have also been produced to permit analysis of the problem. These should take account of:

- [ ] repainting costs for each cycle, bearing in mind that each redecoration may be more expensive on longer cycles owing to greater deterioration
- [ ] the effect cycles have on general rates of deterioration, and the probability of accelerating complete failure of components
- [ ] costs of disruption to productive processes, for both component failures and the decorating operation
- [ ] aesthetic influences that may have hidden costs, e.g. customer service.

Some caution has to be exercised in using published data, as there are always a number of specific localised factors that need to be considered. For example, external painting cycles will be influenced by local climatic and environmental factors, and the orientation and degree of exposure of the building. This may lead to differential weathering, and force the adoption of a different cycle for some elevations. Along with specific user requirements, this may also suggest differing cycles for different buildings within the same estate.

Internal decoration may be influenced by a number of additional factors. These relate not only to the nature of internal materials, but also to statutory requirements in terms of cleanliness and hygiene. The use of the building is also a factor, in terms of hygiene and cleanliness, along with the rate at which decoration deteriorates, and the aspirations of the occupiers. Internal decoration will also be disruptive, so that consequential costs need to be considered. For these reasons the use of locally derived data, if available, is to be preferred.

Forward planning for response maintenance

The striking of an appropriate balance between planned and unplanned work is, in itself, a management decision of some complexity, which can only be based on local experience. The possibility of constructing a highly sophisticated mathematical model exists, but there would be huge data requirements.

There are a number of factors that need to be considered in deciding whether work items go into the planned programme, or become part of contingency planning. These split into those relating to the technical nature of the buildings, and those likely to be determined by the organisational characteristics of the company.

In terms of technical features, clearly the rate of deterioration of components and materials is important, but this cannot be considered in isolation. In the first instance, the building's use will be important because of the way in which this may contribute to deterioration, and also because certain elements may be central to the operation of the building. In purely technical terms it is the predictability of deterioration or failure that is of major importance. Items whose performance is readily predictable offer themselves very conveniently for planned inspection and repair/replacement. Other items will be more susceptible to sudden failure, and although the execution of other maintenance work may help to identify latent potential failures, it is unlikely that this will eliminate all of them.

The important organisational characteristics relate mainly to the speed of information transfer and the service quality required. Increasingly this will be specified via a service level agreement.

It is almost always necessary to adopt, at least in part, a contingency approach to allow for the treatment of those items that may be equally as important as those included in a planned programme, but that may be of an unpredictable nature. This will require, as a matter of policy, the provision of contingency resources in terms of money, labour and materials in order to be able to supply a response. This is vital to safeguard the integrity of the planned programme of work, as well as to provide a service to the occupant (see figure 7.7).

The nature, and size, of this contingency will depend on budgeting constraints, and on the quality of service to be provided. The latter is most directly measured by a response time: that is, the time which elapses between the occurrence of a fault and its repair. In practice it is sometimes taken to mean the time that has elapsed between the maintenance manager receiving notification of the fault and its repair. This approach is not to be recommended, as the speed with which a fault notification reaches the maintenance manager is an essential part of the management system. Mathematical models have also been adopted for this, and although absolute optimisation is difficult and requires accurate data, the principle is quite simple. If the required reaction time is known, and historical data is available, it becomes possible to predict the labour cost of achieving required response times.

The major problem, of course, is in deciding on a reasonable response time. This will depend on the disbenefits that will accrue whilst waiting for a repair to take place. Some of these may be quantifiable in terms of costs, but others will be much more difficult to assess. There have, for instance, been numerous studies carried out regarding the social effects of poor-quality housing maintenance. One of these may be a rise in vandalism, placing increased demands onto already overstretched maintenance budgets.

Another factor that needs to be taken into account is the nature of the defect. Some items may have greater repercussions than others, and although it might reasonably be argued that strategically important items should be part of a planned programme, it would be unrealistic to assume that this always happens. Of particular importance will be items that relate to health and safety, and those that are the subject of statutory requirements.

There are also clear cases where items of a technical nature will require a rapid response. For example, there are defects which if left untreated will become more expensive to remedy, in terms of both direct and indirect costs. A leaking roof, for instance, may not only deteriorate more quickly if left unattended, but the consequential damage will also increase. Although many cases are fairly obvious, there will also be less clear ones of a marginal nature.

At the operational level there are a several ways of dealing with emergency repairs that go some way towards reducing the costs and inefficiencies associated with them. In some organisations emergency repairs are logged in a systematic way. Typically today this will operate through a help desk approach which is possible as an integral part of a facilities management package such as CAFM Explorer.[3] The maintenance team, engaged on their normal planned work, are required to 'pick up' emergency items at the same time as they receive their planned instructions. It has to be accepted that this may cause delays in the planned programme, although these should be allowed for if this approach to emergency maintenance has been formally adopted as a policy.

A similar approach to emergency work is that taken in some hospital estates, where maintenance teams work in one area, for example a ward, only on designated days, at which time they will execute planned work along with any emergency items received.

It is likely that the necessity for a very rapid response tends to load more items into the contingency group in terms of execution, and may additionally put a heavy premium on planned inspections. The requirement for a rapid response capability implies that there will be larger real consequential failure costs.

Making contingency allowances for emergency work is a difficult issue. Theoretically, historical data can be used and indexed for budgeting purposes, but in practice this may not prove to be very accurate. Added to this is the common problem of naturally deteriorating building fabric, which has been exacerbated by under-resourced maintenance activity in previous years, leading to the accumulation of a backlog. This underlines the fact that historical expenditure is often no real guide to future needs.

Bearing in mind the effect that unplanned maintenance work can have on planned programmes, the tendency for backlogs to build up represents part of the vicious circle within which maintenance managers often find themselves.

Planned maintenance programmes

The aims of planned maintenance programmes are extremely diverse, and hence many types of programme will be encountered. The application of the basic principles of planning is of paramount importance. In particular, it is essential to define the objectives of maintenance plans very accurately at the outset, to ensure their relevance and to enable them to be realistically formulated.

These objectives may include all or a combination of the following:

❏ to help ensure that major defects are rectified and that the building fabric is maintained to a defined acceptable, safe and legally correct standard
❏ to sustain the building condition at an acceptable level and prevent undue deterioration of the building fabric and services by preventive means
❏ to preserve the utility of the estate as an asset, and maintain its value
❏ to maintain the engineering and utility services in an optimum condition to safeguard the environmental conditions of the building, and hence its productive capacity
❏ by effective planning, ensure that maintenance is conducted, over a number of years, in a sensible sequence that reflects a careful consideration of priorities
❏ by proper planning, ensure that maintenance operations are carried out in the most effective way to ensure that best value for money is being obtained and the best use is being made of scarce resources
❏ to provide a tool for financial management, in particular budgetary control, and to assist maintenance managers in bidding for financial resources
❏ as part of a broader facilities management scenario, to assist management to relate programmed repairs and maintenance to other demands and alternatives, such as refurbishment, redevelopment or changes in leasing policy.

The characteristics of maintenance work make accurate and comprehensive long-term predictions rather difficult. It is therefore necessary to define carefully what is realistically possible, and have an explicit recognition of levels of uncertainty. Because of this, all programmes will need to have built into them some flexibility to permit modification and updating in order to ensure their continuing relevance.

Failure to understand uncertainty and risk leads to a tendency to persist in the adoption of unreal assumptions that fail to recognise reality, and this has undoubtedly led to the misuse and abuse of modern planning techniques.

Maintenance has its own characteristics that further complicate the planning process.

❑ The work is characterised by a large number of small jobs, and attempts to programme individual jobs in minute detail, over more than the short term, are clearly not realistic.

❑ The widely dispersed nature of much of the work is a major factor to be taken into account when planning, as it has a major impact on efficiency and economy.

❑ Individual jobs are often simple in nature and in terms of sequencing, but there is more of a need to consider logistics, rather than detailed working methods.

❑ A large proportion of very small jobs may require the presence of a number of trades, the co-ordination of which is difficult. This makes the achievement of continuity of work for individual trades difficult.

❑ The work content of a maintenance item may be uncertain when an order or instruction is given, and the extent of a repair may only reveal itself when the building fabric is opened up.

❑ The adoption of a conscious policy for emergency items to be attended to, as part of a planned visit to a location, underlines the need for flexibility.

❑ The most carefully constructed work programmes are subject to disruption from a number of potential causes:
  ○ withdrawal of resources to deal with emergency work
  ○ climatic conditions
  ○ access problems
  ○ budgetary setbacks.

❑ Emergency repairs present very specific problems owing to their unpredictable nature, often allied with the need for a rapid response, thus creating very short lead times.

In view of these difficulties, programmes are formulated at a number of different levels and each of these may be used in a different way. In general, the following categories of maintenance programme can be identified:

❑ long term   quinquennial or longer
❑ medium term   annual
❑ short term   monthly, weekly or even daily

In terms of management information these programmes may not all always be necessary, and three levels can be considered in this respect.

(1) A programme may be required for financial management purposes only, in which case it is likely to be a long-term one with, perhaps, subsidiary medium-term ones. In basic terms it will comprise maintenance costings, on a building-by-building basis, perhaps with some prioritising system built in to aid decision-making in the event of budgetary constraints.

(2) A further programme may be required for executive management purposes, and in this case a more detailed data input will be required in terms of a breakdown

of each building's maintenance needs, probably on an elemental basis, accompanied by costings. It will also require the allocation of a timescale for operations and costings. This type of programme is most likely to be a medium-term one, but with shorter-term updates or breakdowns.

(3)   For operational and works monitoring purposes, shorter-term programmes may be required, and these will need much more detailed information about the tasks to be performed.

Consideration of the above programmes helps to highlight several things.

❑   The various types of programme are linked and one may follow from another through a top-down or bottom-up approach.
❑   There are no hard and fast rules about timescales for programmes; it is only necessary to follow the principles carefully, and to match these to the needs of the organisation.
❑   Typically, in any maintenance organisation there is a maintenance planning system that represents a composite approach.
❑   For an organisation introducing planned maintenance for the first time, the cost of acquiring data is a major task, and it may well be prudent to start at (1) and build up to (3) as the data is collected.
❑   It is worth bearing in mind that the data collected will have a variety of uses in addition to planning purposes. Indeed, it is the case with contemporary computer-based planning systems that the planned maintenance programme is part of an integrated management information system.

Long-term programming

The objective of long-term programming is not to set down detailed task data or precise dates for individual operations. This would clearly be unrealistic, especially as the objectives of a long-term programme do not require this type of detail. The application of the exception principle is of vital importance, and long-term programmes should not be overloaded with spurious detail, which will only serve to confuse the real issues. To produce a long-term maintenance programme may only require a systematised broad survey of the estate, such as that commonly used in the NHS.[4] In this way a picture of the overall problem can be obtained for the purposes of developing and putting forward a strategic plan of action, or to evaluate alternative strategies.

The period of study for this type of exercise will depend on the characteristics of the estate and the objectives of the organisation. There is no generally accepted industry standard. Fifteen-year programmes have been proposed, but in themselves are unlikely to be realistic. However, there is a tendency to work in five-year blocks, so that a ten-year programme may be produced where one level of certainty can be placed on the first five years of the programme and less on the second five.

As the programme progresses, it should be capable of revision and updating, leading to a firming up of longer-term predictions as time goes by. There is now a distinct tendency for many organisations to utilise this type of rolling programme.

Longer-term programmes can be produced for a variety of purposes.

(1) The determination of the expenditure required for maintenance over a period of time, in order to put and keep the building stock in an acceptable condition. Where there is a maintenance backlog problem, it is useful to separate the backlog work from future requirements when drawing up the programme. This separation will often emerge in any case, if a prioritisation system is incorporated.

(2) Possibly linked to (1), or stemming from it, is a long-term programme that can be used to plan expenditure streams in the most effective way according to circumstances. An obvious example is where expenditure has to be planned within the constraints of available finance. This again encourages the use of a priority coding system.

(3) Major repair and/or renewal proposals require careful forward planning, to be consistent with the financial resources available, and to ensure that their timing has the minimum disruptive effect on the organisation. Major repair programmes are often a special case, in that they are more likely to be the result of a rigorous analysis of a specific problem, and usually lend themselves to a more precise definition of both the cost and the work content. There is often some flexibility in the planning of major repair programmes, in that they rarely contain emergency items. This fact can be very useful, as it may permit an adjustment to a programme where, perhaps, some major unpredicted failure occurs and swallows scarce resources in one financial year. An example of this is illustrated in figure 7.10.

(4) Long-term forward planning is useful for budgeting for financial resources. However, it has to be borne in mind that other resources, particularly labour and materials, have to be available, broadly at the right time. Whilst this can be achieved to some extent by long-term planning, detailed analysis of material and labour requirements is more effectively carried out as a medium-term exercise.

(5) The broader dictates of facilities management will make use of long-term maintenance plans, which will need to be integrated with other longer-term ventures such as those relating to new capital projects, refurbishment, demolition and changes in the use of the building stock.

Several elements of the programme may be readily predictable, but as the timescale increases, so the level of uncertainty becomes greater. This is frequently the case with cost predictions. Figure 7.10 shows an example of part of a long-term rolling programme, where a planned programme is being instituted for the first time. During the first year there are some unforeseen events, and changed conditions revealed by more detailed inspections. The programme is adjusted accordingly.

It is also possible that the need for a completely unforeseen renewal programme becomes apparent through the first year, and to accommodate it the programme would require further modification and updating.

A further example of part of a long-term programme, based on financial predictions, is shown in figure 7.11.

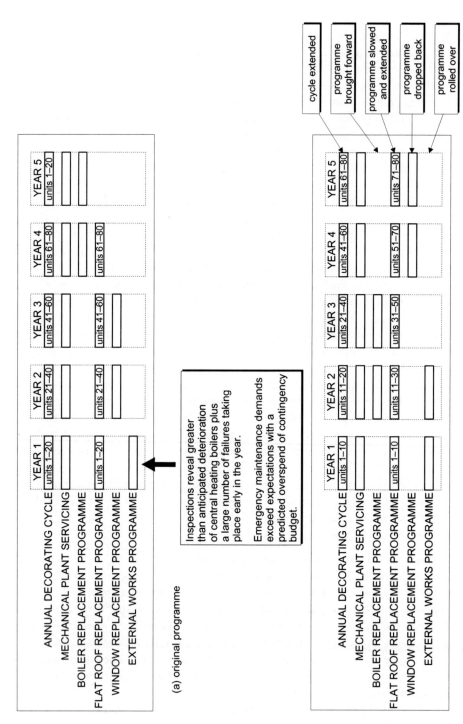

Figure 7.10 Adjustment of repair programmes over a five–year period.

LONG-TERM MAINTENANCE PLAN FOR MILL LANE TEACHING BLOCK
FIVE-YEAR BUDGET FOR THIS BUILDING £150 000 (£142 250 ALLOCATED)
PROVISIONAL 15-YEAR EXPENDITURE CURRENTLY £325 000 (£300 750 ALLOCATED)
CONTINGENCY CENTRALLY CONTROLLED

ITEM OF WORK	PRIORITY CODE	1996	1997	1998	1999	2000	2001–2005	2006–2010
cyclical maintenance								
decoration		13 000	12 000	13 000	13 500	14 000	15 000	12 000
cleaning		3 000	13 000	1 500	1 500	1 500	50 000	50 000
external work		1 500	2 000	1 000	1 000	1 250	5 000	5 000
plant servicing		1 000	500	1 000	500	2 000	6 500	7 000
inspections			500				4 000	4 000
remedial works								
repoint verges	high	1 000						
repair paths and steps	medium		3 000					
repair damaged plaster	medium		1 000					
remedial work to internal joinery	low		2 000					
major programmes								
boiler replacement	high		20 000		5 000			
window replacement	low			5 000				
flat roof renewal	medium			5 000	2 000			
replace gutters and downpipes	low							
TOTAL EXPENDITURE		19 500	54 000	26 500	23 500	18 750	80 500	78 000

Figure 7.11 Example of part of a long-term maintenance programme.

Intermediate or medium-term programmes

These will normally be annually based, although updating at intermediate periods should be accepted as inevitable. The intervals at which this takes place will vary, depending on where the maintenance management system fits within the overall organisation. In many cases the reporting and control mechanisms will be driven by accounting procedures, requiring monthly reporting and review to be adopted.

At the outset, annual programmes will be set within the framework provided by longer-term planning. However, at the formulation stage there may well be specific needs that have been identified for inclusion in the programme for that year, leading to revisions in long-term plans. Information will therefore flow in both directions between medium- and long-term programmes, and this underlines their essential interdependence. At the level of the medium-term timescale there will be less uncertainty, and it would logically be expected that a greater level of detail would be built into them, so that they become of much more benefit for operational purposes.

Several important uses of medium-term programmes can be identified in this context.

(1) They assist managers to determine the allocation of the annual maintenance budget, and to plan the way in which this expenditure will occur in the most effective and appropriate way, so that good short-term financial planning is achieved.

(2) From the operational point of view, money is one resource whose usage has to be controlled, and hence needs to be planned. The annual programme is the essential tool for planning the optimum usage of the resources that cost money, namely, labour and materials. If labour is in house, then use may be made of histograms and other similar techniques for the purpose of resource optimisation exercises.

(3) If the work is to be executed by contractors, then the planning exercise is still just as relevant to ensure that a proper time period is available for the preparation of contract documentation and contractor selection.

Where a number of term contractors are employed, their work will need to be balanced and apportioned effectively so that:
- ❑   no contractor becomes overloaded
- ❑   the right work is allocated to the right contractor on the basis of his strengths and weaknesses, wherever possible
- ❑   geographical issues are taken into account.

(4) In terms of materials use, the annual programme should be able to provide indications of the quantities of key materials and components that will be required during the year. This will enable a buying and storage strategy to be developed, ensuring that materials and components are always available as required, and that they are obtained on the best available terms.

(5) The annual maintenance programme should enable maintenance work to be timed as conveniently as possible, so as to minimise disruption within the client organisation and reduce real costs.

Examples of parts of an annual programme are shown in figures 7.12 and 7.13, with the work broken down as follows:

❑ individual jobs
❑ division into major and routine jobs, with an allowance for emergency work
❑ division into that to be executed by direct labour and that by contract labour
❑ expenditure estimated and allocated to each job, with further subdivision into labour, plant and material costs, so as to help medium-term resource planning.

The preparation of the annual programme can be divided into the following steps, which are broadly in line with the principles outlined at the beginning of this chapter.

(1) Identification of items to be included in the programme

❑ Items brought forward from the long-term programme, although they may need verification through inspection.
❑ Items identified from an inspection, which are required to be carried out in the coming year. If these items are not in the longer-term programme then the implications may need to be assessed for the long term. If the long-term planning has been carefully carried out, there should be a contingency item from which these items can be funded.
❑ During inspections, requests for items of work may arise from the occupants. In some organisations it is the practice to ask occupants to nominate items prior to a regular inspection, which enables some judgements to be made in the context of budgetary and resource constraints.
❑ An allowance for unforeseen maintenance items.
❑ Routine day-to-day maintenance items, of which the greatest contributor will probably be cleaning.

(2) Identification of work content and costs
If a good data management system exists, then feedback from previous work is the major tool to be used for this purpose. Decisions may have to be made as to which items are to be executed by direct labour, and which by contractors. A policy framework for this may already exist in the organisation.

The breakdown of the work content of the item enables appropriate material and labour inputs to be identified, and hence costs. There may, at this stage, be a number of detailed planning exercises carried out, such as for labour planning.

(3) Determining the sequence of work
The use of critical path networks is rarely relevant except for the biggest maintenance exercises. There will, however, often be a substantial logistics problem to solve, and thus the possible use of operational research techniques should not be excluded.

It is in this aspect of planning that computer-based systems come into their own, because of their ability to manipulate data very quickly. This enables the planner to analyse and evaluate a variety of working methods and sequences, and thus make

MILL LANE TEACHING BLOCK – ANNUAL MAINTENANCE EXPENDITURE BREAKDOWN 1997
TOTAL BUDGETED EXPENDITURE £56 000 INCLUDING PROVISION FOR UNPLANNED WORK

ITEM OF WORK	PRIORITY CODE	TOTAL EST. COST	CONTRACT	DIRECT LABOUR		
				Materials	Labour	Overheads
routine maintenance						
decoration		12 000	12 000			
cleaning		13 000		3 000	7 000	3 000
external work		2 000		300	1 500	200
plant servicing		500		100	350	50
inspections		500			450	50
individual tasks						
boiler replacement	high	20 000	20 000			
repair paths and steps	medium	3 000		700	2 000	300
repair damaged plaster	medium	1 000		200	700	100
user requests and emergencies		4 000				
TOTAL EXPENDITURE		56 000	32 000	4 300	12 000	3 700

Figure 7.12 Budgeted annual expenditure.

MILL LANE TEACHING BLOCK – ANNUAL MAINTENANCE PLAN 1997

ITEM OF WORK	PRIORITY CODE	EST. COST	EXECUTION	JAN	FEB	MAR	[APR]	MAY	[JUNE]	JULY	AUG	[SEPT]	OCT	NOV	DEC
routine maintenance															
decoration		12 000	contract				4 000			2 000	4 000	2 000			
cleaning		13 000	direct lab.	1 000	1 000	1 000	1 500	1 000	1 000	1 000	1 500	1 000	1 000	1 000	1 000
external work		2 000	direct lab.									2 000			
plant servicing		500	direct lab.							500					
inspections		500	direct lab.												500
individual tasks															
boiler replacement	high	20 000	contract							5 000	10 000	5 000			
repair path sand steps	medium	3 000	direct lab.							3 000					
repair damaged plaster	medium	1 000	direct lab.				1 000								
user requests and emergencies		4 000	direct lab.	450	450	450	450	350	250	150	150	150	300	400	450

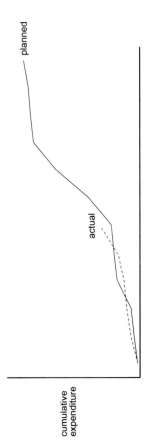

cumulative expenditure

planned

actual

Figure 7.13 Budgeted annual expenditure and maintenance programme.

choices that will optimise a programme. Automated data management also permits the application of a range of decision-making aids.

(4)   Provision of a controlling mechanism

The programme that is set up through the preceding process now becomes a model of the work of the maintenance department in the medium term, and is thus an essential component of a management control system. It is necessary for actual achievements to be compared against predicted, in order to assess progress and prompt management to take corrective action. One of the first prerequisites here is a feedback mechanism to collect appropriate information by which progress can be measured.

In any planning system there are numerous ways in which progress can be recorded but this is now likely to be through a computer-based system, where data can be more easily and quickly processed. Such systems also encourage the adoption of a flexible approach to planning, as continuous updating is much more easily effected.

Short-term programmes

When short-term planning is considered, more detailed aspects become of importance for two reasons:

❑   Analysis of performance, and the provision of feedback data is the essential source of information for future planning exercises, and for this to be facilitated short-term programmes require a detailed basis against which data can be collected, compared and analysed.

❑   For operational purposes, and day-to-day management control, it is important that all inputs to a maintenance operation are properly identified and stated.

At the beginning of each year, monthly programmes may be produced by simply breaking down the annual programme into 12 workloads. Monthly programmes produced in this way will almost certainly have limited value and, bearing in mind the principle of ensuring that planning should be relevant and realistic, caution should be exercised. However, figure 7.14 illustrates typical examples of some monthly planning exercises.

There may well be a case for taking a forward view in more detail, perhaps on a quarterly basis, to help good resource management but, realistically, a prediction of a month ahead seems reasonable. Quarterly programmes may, nevertheless, have some benefit when a large proportion of contract labour is being used, as this is most likely to be for predictable work and the responsibility for resource management shifts to the contractor. The client, under certain circumstances and types of contract, may require the contractor to produce a programme for the work he is to execute.

When a high proportion of direct labour is being employed, short-term planning may take place at a variety of levels, and the possibility of monthly programmes having to match accounting cycles, for example, has already been mentioned.

At the operational level there are likely to be weekly programmes, and sometimes daily ones. When a maintenance team collects its work allocation for a day, this can be

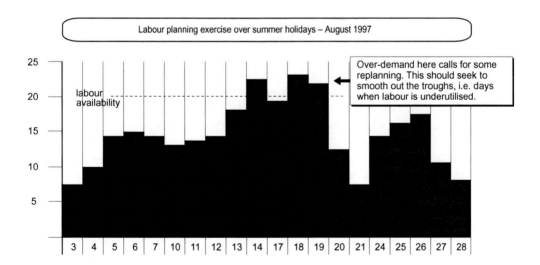

MILL LANE TEACHING BLOCK – major works – summer 1997 short-term programme

ITEM	July				August					September			
	6	13	20	27	3	10	17	24	31	7	14	21	28
Decoration													
Repair path sand steps													
External works													
Plant service													
Boiler replacement													
Prepare new boiler house													
Re-route existing services													
Install new boiler													
Make connections													
Commission													
Builders work													
Remove old boilers													
Refurbish old boiler house													

(a) Programme of works over summer period in one building.

Labour planning exercise over summer holidays – August 1997

labour availability

Over-demand here calls for some replanning. This should seek to smooth out the troughs, i.e. days when labour is underutilised.

| 3 | 4 | 5 | 6 | 7 | 10 | 11 | 12 | 13 | 14 | 17 | 18 | 19 | 20 | 21 | 24 | 25 | 26 | 27 | 28 |

(b) Assessment of labour requirements for whole estate during August 1997.

Figure 7.14 Typical short-term programmes.

thought of as a short-term programme of work. When maintenance operatives collect a works order, execute the work and complete the works order, they are in fact working within a planning system.

Many of the optional contributions of the annual programme become of prime importance in the shorter term. This is particularly the case for manpower planning. Even though the annual programme and the short-term programme tend to complement each other, both are necessary to make best use of the workforce. In terms of operational management, the detailed programming of work depends on the extent of the direct labour contribution, and the need to ensure a smooth programme of work for it, which also reduces travelling time to a minimum.

The short-term programme, in a sense, puts into action the forward planning of the annual programme, and feedback to it will enable monitoring and control for the whole year.

The three levels of planning identified above are evidently not mutually exclusive, and the boundaries between them are indistinct. All three should be viewed as integrated parts of a comprehensive planning system.

Presentation

The most familiar presentation format for a programme is the bar chart, sometimes termed a Gantt chart, after one of the early twentieth-century pioneers of management. The bar chart has the advantage of simple visual communication, and can be tailored to include many items of information. The examples shown in figures 7.2, 7.3 and 7.4, for a new-build project, are indicative of its usefulness.

The bar chart has been used for many years as a technique for new-build construction work. Its main deficiency is that on its own it does not indicate the relationship between activities. For this reason there has been, over a number of years, a growing impetus for the use of critical path networks on the grounds that they remedy this. However, in terms of presentation they are not particularly appropriate, and to be really useful it is necessary to generate a bar chart from the network.

The benefits of a network-based approach are now being more fully realised, with the availability of cheaper and more user-friendly hardware and software. There is no doubt that this is the case for a great number of more sophisticated planning techniques, with the added advantage of enormous flexibility in the ways in which information can be presented.

Computerised management systems are able to present information in tabular forms, as schedules and histograms, and also in a graphical form. The majority of good networking packages also have the ability to export to or import data from spreadsheets and databases, so that they become part of an integrated information management system.

Networking techniques undoubtedly have their application in planning for repair and maintenance, but this is limited to larger one-off projects of a more complex nature. The majority of maintenance activity, however, is represented by a large number of small activities, many of a routine nature.

Management control

There are several components that need to be present for a control system to be effective:

❑ established benchmarks or performance standards
❑ measurement of performance
❑ comparison of performance against that required
❑ corrective action.

A fifth component that could be added to these is the presence of a satisfactory information system that communicates, collects and analyses data, and channels it in the appropriate manner.[5]

A basic question that has to be addressed at the outset concerns what has to be controlled, and in the context of maintenance operations this will be quality, time and money.

Control methods, in very simple terms, can normally be categorised into:

❑ those focusing on physical values of measurement such as quality, but also perhaps encompassing output performance
❑ those focusing on financial values and which are intrinsically linked to time management.

The normal mechanism for the latter is a budget. In the management of maintenance this will be an expenditure statement, and the process of control will require the monitoring of costs.

The nature of costs

Costs can be split into four elements:

❑ labour
❑ materials
❑ plant
❑ overheads[6]

For new-build construction activities the separation of plant may be essential, but in many cases, when its use is of a general, rather than a directly attributable nature, plant costs are subsumed into overheads.

A direct cost is a cost that can be identified with, and allocated to, an operation. For example, materials and labour expended on a repair item are clearly direct costs, as they can be readily attributable, provided there is a proper information system. All other costs are termed indirect costs, for example supervision, or the use of small plant items not directly attributable to individual items of work.

Overheads may be defined as the aggregate of indirect costs. Proper financial management and control requires that overheads be accounted for in some way. In the normal maintenance department there are two categories of overhead to be considered.

In the first instance, there are the indirect costs directly associated with the maintenance department, all of which can be attributed to that department. There will also be indirect costs associated with the overall running of the company, which cannot be allocated to individual parts. In this case a method of apportionment has to be used.

A maintenance department budget will, therefore, include a portion of the organisational overheads, its own attributable indirect costs and the direct cost of executing the work. Note also that if contract labour is being used for execution, then the price paid for that service will include contractors' overheads.

Costs may vary with output, in which case they are termed 'variable costs'. Materials and labour expended in executing maintenance work are obvious examples. Both direct and indirect costs may be variable.

Fixed costs do not vary with output, but will vary with the passage of time. These may also be either direct or indirect costs. It is important to appreciate that fixed costs may only remain fixed over a certain output range. For example, at a given rate of maintenance expenditure, the cost of supervision is a fixed cost. If expenditure is halved or doubled then supervision requirements will change, leading to a variation in a fixed cost.

In reality many costs may fall between the two and be termed 'semi-variable'.

Absorption and marginal costing

There are two ways of associating indirect costs with production. In absorption costing, all costs – direct and indirect – are charged to the unit of production, so that a unit cost has absorbed a portion of the fixed costs. If output drops, then there are fewer units produced to share these costs, so they will all need to take an extra share: that is, unit costs increase. The converse, of course, is true and, if output rises, unit costs drop. In theory, therefore, the unit cost of one more unit is slightly less than the cost of the previous one.

In a marginal costing approach, only variable or marginal costs are charged to the process. In a commercial venture, fixed costs are written off against profit in the period in which they occur. The cost of one more unit of production is simply the variable cost.

Both approaches have their relevant applications, but use of the most appropriate is extremely important because of the influence it may have on management decision-making. For example, a maintenance contractor may choose to take the former approach in tendering for maintenance work. A direct labour organisation, on the other hand, may prefer to operate on a marginal-cost basis when deciding whether or not it can carry out an extra repair.

Standard costing

An important, and useful, concept is that of standard costing. This is a predetermined cost, derived from previous records, having regard to normal standards of efficiency, which is used as the basis for price-fixing and costing. Schedules, or rates for maintenance work, or historical data, internally generated within the organisation, may be used as the basis for deriving standard costs.

Standard costs can be particularly valuable for budgeting purposes and other forward financial planning exercises.

Budgets and budgetary control

A budget is defined as a financial, and/or quantitative statement, prepared and approved prior to a defined period of time, of the policy to be pursued during that period, for the purpose of attaining a given objective. It may include income, expenditure and the employment of capital.

Budgetary control is defined as the establishment of budgets, relating the responsibilities of executives to the requirements of a policy, and the continuous comparison of actual with budgeted results, either to secure by individual action the objective of that policy, or to provide a basis for its revision.

In practical terms, budgets can be considered to be a planning function that produces management information. Budgetary control is, however, both a planning and an executive function, in that the term 'control' implies action.

Within an organisation, the formulation of budgets is a complex process (figure 7.15) and requires patience and high-quality management information. Sensible costing data is obviously essential, as the budget is fundamentally a financial forecast.

Inevitably, however, the maintenance department will find itself in difficulties. Theoretically, given an up-to-date performance model of the estate and good cost data, a maintenance manager may produce a budget, based on his medium- and long-term maintenance programmes. It would be naive in the extreme, however, to assume that the budget will ultimately be driven by technical need. The maintenance manager may present a budget based on his needs, but will often be forced to accept less, so that his budget for the year is driven by financial realities.

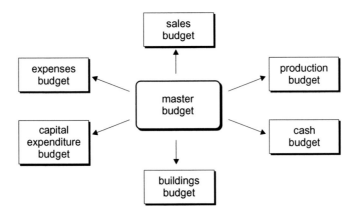

Figure 7.15 Relationship of section budgets to master budget.

Structure of a maintenance budget

In the same way that master budgets are sectionalised, departmental budgets are subdivided under so-called budget headings. Notwithstanding this, the relationship of departmental budgets to the overall budget is important, and should be accompanied by a policy statement.

The breakdown of the maintenance budget can be carried out in a number of ways:

- ❑   by type of cost
- ❑   by type of work, e.g. a trade breakdown
- ❑   by building or parts of a building
- ❑   on an elemental basis
- ❑   by who will execute the work.

In reality, combinations of these will be required to provide a proper control tool. However, of most significance will be a classification of the degree of importance associated with each operation.

Like any other programme, budgets should be flexible and dynamic, and there must be an acceptance that items will be dropped. Most maintenance budgets will also include a contingency for unplannable items. On the basis that this contingency may not be totally used, the budget may also include discretionary items, perhaps ranked in terms of priority, which can be brought in. Examples of the ways in which this might operate were shown earlier.

The budget represents an incremental financial statement over a fixed period, normally a financial year. For proper control purposes, there must also be a forecast of the pattern of expenditure. If there are properly formulated maintenance programmes, this information should be derived quite logically from it (figure 7.13).

A number of appendices to the budget statement may be useful. Detailed cost breakdowns and inspection reports are an example. However, these are more properly a function of a good information system, of which the budget is a part.

Cost reporting

The maintenance information system must be designed to report actual costs against what was planned. This has to be carried out at regular intervals, normally monthly. A typical outline report is shown in figure 7.16.

The term 'actual cost' needs to be treated with some caution. Management information must be produced as quickly as possible, and it may be better to sacrifice accuracy in order to satisfy this need. Adjustments may, therefore, need to be made in succeeding periods. The cost variations reported may be further analysed by causes, to help management decision-making. If standard costing has been used, for example, it may be useful to report variances in financial reports. Statistical techniques may be of value here, in that some variance from a standard cost would be expected. It is variances beyond what would be expected that are of importance.[7]

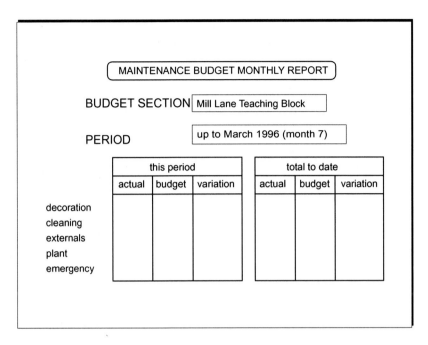

Figure 7.16 Simple budget report.

Forecasting

Like any programme, budgets will be subjected to change and adjustment. Periodic reports should, therefore, include a predicted financial outcome, which is essentially a revised financial programme. This will need to take account of a number of different costs.

(1) Costs that are budgeted from the original statement.
(2) Actual costs, which may be subjected to revision in the next period.
(3) Committed costs, which represent the consequences of irreversible decisions. The original budget may of course have included elements of this type. Also of importance, in interim reports, are costs required to complete tasks already embarked upon, which may differ from the original prediction.
(4) Variations from budget figures in the predicted costs of future planned items, due to additional information coming to light.

The forecasting exercise will produce an up-to-date prediction of the financial outcome, and this is more useful for management control than a simple actual versus planned statement.

Corrective action

Depending on the causes of variances from the budget, and also the quality of associated management information produced, there are a number of possible scenarios.

If the budget is being, or is likely to, overrun then decisions will need to be made to take some work out of the programme, or perhaps adopt some economies in non-programmed work. However, this only represents an immediate response, and for further management action some analysis of the cause of the problem needs to take place.

The variance may be due to:

❏ inaccurate cost forecasting, in which case remedial action needs to be taken for future forecasting exercises
❏ actual costs deviating from predictions owing to inefficient working practices
❏ inefficient organisation and planning of the work
❏ variations in the scope of the work, which may be unavoidable or due to a poor inspection regime.

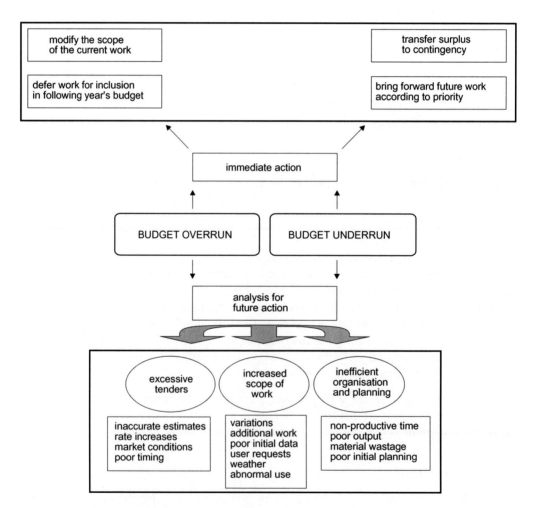

Figure 7.17 Analysis of variations from budget.

There are clearly a large number of possible causes, but these cannot be identified properly unless the appropriate data is collected and presented. A systematic approach to the analysis of the variance is suggested in figure 7.17.

References

(1) NBA Construction Consultants (1985) *Maintenance Cycles and Life Expectancies of Building Components: A guide to data and sources.* HMSO, London.
(2) HAPM (1999) *Component Life Manual*(CD ROM). Spon Press, London.
(3) www.cafmExplorer.com
(4) Spedding, A. (ed.) (1987) *Building Maintenance and Economics – Transactions on the Research and Development Conference on the Management and Economics of Maintenance of Built Assets. Management of built assets – value for money.* E. and F. Spon, London.
(5) Cole, G.A. (2004) *Management, Theory and Practice,* 6th edn. DPP, London.
(6) Davies, D. (1994) *Finance and Accounting for Managers,* 2nd edn. Institute of Personnel Management, London.
(7) Dyson, J.R. (2004) *Accounting for Non-Accounting Students,* 6th edn. Pitman, London.

Chapter 8
Maintenance Contracts

Introduction

The general legal requirements for the formation of a maintenance contract apply in the same way as they do for any other building contract. The major issue for the maintenance manager is to select the type and form of contract that is most appropriate for their purposes.

The basis of any contract is the offer by the Contractor to carry out work for a sum of money, and the acceptance of that offer by the Employer. Although there is no legal necessity for the contract to be in writing, it is generally accepted that this is highly desirable from the administrative point of view. It is also accepted practice that supporting documentation, termed 'contract documents', is prepared, to assist in the proper operation of the work covered by the contract.

There are several standard approaches to putting together contractual packages, each designed to fit a particular set of circumstances. Over the years, there has been a proliferation of standard forms, and much debate and research has taken place concerning the appropriateness of each for new-build, repair, maintenance and refurbishment work.

The main differences between the various types of contract available relate to the methods of evaluating work, and the degree of financial risk to be borne by the parties. This is reflected in differing arrangements for pricing, valuing and paying for the work, and in the supporting documentation. The way in which these aspects are handled is clearly set out in standard clauses, forming part of the standard form of contract being used.

Building contracts generally

Fixed price contracts

These are contracts where the price is agreed and fixed before the contract is signed. The agreed price is paid irrespective of the builder's actual costs, subject to a set of Contract Conditions allowing them to recover additional costs caused by agreed circumstances, such as variations or alterations to the contract documents on which their price was based.

These types of contract can be considered to be risk-bearing, a degree of risk being apportioned to each party. Fixed price contracts are divided into two types.

(1)   Lump sum contracts

With these contracts, the Contractor agrees to execute the whole of the work for a stated 'lump sum', which is based on firm quantities, specifications and drawings. There are two derivatives: firm price contracts, which do not allow for contract prices to be adjusted for fluctuations in market prices of the Contractor's costs; and fluctuating price contracts, which provide for the use of a formula for the adjustment of prices to take account of these movements.

Lump sum contracts presuppose that there is sufficient information available, prior to the tender stage, to permit accurate preparation of a firm bill of quantities and specification, and that therefore the client's level of risk is minimised through the submission of a firm price for the job. Financial administration of the project is readily facilitated, as a firm basis exists from the outset.

However, in many cases contract documentation is founded on imprecise information, and if this happens the presumed advantages may not accrue. In consequence, this form of contract, together with traditional competitive tendering methods, has come under increasingly critical scrutiny for new-build work.

Except for major renewal schemes, where firm positive documentation can be prepared, this type of contract is not generally appropriate for maintenance work.

(2)   Measure and value contracts

In these forms of contract, the Contractor agrees to execute the work, at prices fixed in advance, for units of work to be measured later. These may again be firm or fluctuating, depending on arrangements for coping with market movements in Contractor's costs. In some instances the contract documents may include a set of approximate quantities, and this provides the client with a better idea of the extent of his final outlay. However, the indiscriminate use of approximate quantities should be discouraged, as they are often of little more administrative use than a Schedule of Rates, and may present a misleading picture to the client.

Another variant is the so-called schedule contract, which is used where the work details are too vague to permit the preparation of fully detailed information at the time of its commencement. The schedule, listing all the items of labour and materials that are expected to be used, may be prepared specifically for the job in hand, or be a standard schedule prepared to cover a range of jobs.

The schedule may be unpriced, in which case the Contractor is required to enter prices against the schedule items; this is then used as a means of selecting a Contractor. Standard priced schedules are also used, where the Contractor tenders by giving a percentage on or off the standard rate. For example, the former Property Services Agency (PSA) of the DoE made great use of a standard pre-priced schedule. The work is measured at completion, or interim stages, by using the relevant rates.

Although a convenient way of providing in advance for the pricing of work of uncertain content, standard priced schedules lack estimating accuracy. Contractors tendering are required to state their offer in terms of a percentage on or off the schedule rates as a whole, which assumes that an overall percentage adjustment will bring them into line with those normally charged by the Contractor. This is unlikely to be the case for a number of reasons.

❑   There is little uniformity among Contractors regarding the pricing of individual items of work, and inevitably some of the schedule rates will be higher and some lower than a given Contractor would normally use.
❑   The mix of items for a particular job may well exaggerate the effects of such differences in pricing patterns. For example, a job may consist largely of those items for which the standard rates are higher than the Contractor's usual ones and the addition of a percentage would only serve to heighten such discrepancies.
❑   The final cost will depend upon both the quantity of and the rate for each item in the work executed. If one accepts that some of the schedule rates are high and some low, then the relative quantities of each influence both the outcome and cost predictions.
❑   The schedule rates are essentially averages, and may not reflect the particular conditions under which the relevant work items have to be carried out.
❑   Standard rates tend to need updating at more frequent intervals than actually occurs in practice.

In order to arrive at a realistic estimate, the tenderer should estimate the likely proportions of schedule items in the job at hand, and then determine a percentage adjustment that will equate the cost based on the Schedule of Rates to the cost he would obtain if using his normal rates. However, schedule contracts do fulfil a useful role for many types of work, notably maintenance.

Cost reimbursement contracts

In these types of contract the Contractor is reimbursed for the actual prime costs of labour, materials and plant used, plus either a previously agreed percentage, or a fixed fee, to reimburse him for his management costs, overheads and profit. In its pure form, the disadvantage of this type of contract is the absence of an incentive for the Contractor to keep his costs down. Its use is therefore restricted to small or urgent jobs where the necessity to execute work very rapidly provides insufficient time to produce precise documentation, and the degree of risk to the client can be justified.

This type of contract requires a reputable Contractor who is known to the client. Indeed, if the client is a good and regular customer, this should help to limit the risk, provided there is a proper checking mechanism at the end of the job.

The low level of documentation places an increased burden on supervision, and requires a very good communication system to be in place during construction. The client's representative, in particular, has a much more demanding job in protecting the client's interests, and a larger measure of interference with construction methods is justified, and should be expected by the Contractor.

The competition element is limited, and depends, in substantive terms, on the Contractor's percentage add-on. This in itself may not always be the best guide, as a higher percentage add-on may reflect a more efficient attitude to management of the project, which may lead to a smaller final cost to the client. The converse, of course, may also be true.

The best method of selection is probably on the basis of reputation, which may be fine if the client regularly lets this type of contract to a limited number of Contractors, but is clearly problematical when the client has no prior experience upon which to draw.

Another factor that must be borne in mind is the checking procedure to ensure that the client obtains what he has paid for, and also the scrutiny of the final account for future reference. This is a notoriously difficult thing to do, especially when it comes to assessing the number of labour hours included. The situation may be helped by requiring the Contractor to submit records of resource inputs, at regular stages throughout the execution of the work, to permit ongoing checking and control.

Use has been made of so-called controlled Daywork, where all jobs are pre-estimated prior to placing the order. If the Contractor's account exceeds the estimate by more than an agreed percentage, the reasons are investigated. This method has the merit of providing the client with a firmer idea of his outlay, although the pre-contract period may be somewhat lengthened.

Latterly, increased use has been made of cost reimbursement contracts for large projects, let on a management contracting basis. This is generally of little relevance to maintenance contracts but there are some developments that are interesting, in particular the use of target cost contracts, where an initial prime cost target is agreed, along with a management fee and an adjustment percentage. Any variation from the target cost then leads to recalculation of the management fee in accordance with the adjustment percentage. The simple example shown below illustrates the principle.

	Initial prime cost target	£60 000
	Management fee	£5 000
	Adjustment percentage	25%
(a)	Prime cost at completion	£62 000
	Contractor is paid:	
	Prime cost	£62 000
	Management fee	£5 000
	Less an adjustment of 25% of excess of £2 000	£(500)
	Total payment	£66 500
(b)	Prime cost at completion	58 000

Contractor is paid:	
Prime cost	£58 000
Management fee	£5 000
Plus an adjustment of 25% of saving of £2 000	£500
Total payment	£63 500

Thus, in this form of a cost reimbursement project, there is both incentive and penalty for the Contractor, and an adjustment process, which shares costs and benefits between Contractor and client.

Fixed price maintenance projects

This approach was adopted several years ago, in an attempt to reduce the large amounts of paperwork involved in executing types of contract in which the administrative costs for small projects may conceivably exceed construction costs.

The method involves agreeing a lump sum, based on analysis of maintenance records, with a Contractor for undertaking a range of recurring works of a similar nature, to a specified group of buildings, over an agreed period of time. The Contractor then agrees to carry out all the work of the specified types for the agreed contract sum. Tendering is through the Contractor quoting a percentage on or off the given lump sum. The stipulated time periods are normally a year, and the Contractor is paid one-twelfth of the lump sum each month.

In theory, this approach makes large savings in administrative costs, and payments are simply made on a monthly basis. The checking necessary can be through regular site inspections to ascertain that the work is being executed properly.

It has also been argued that there is in-built quality control, as failure of a Contractor to execute work properly will manifest itself at an early stage through a call for the work to be done again. This is by no means certain, unless the same Contractors are regularly employed for the same tasks on the relevant buildings. However, if the system is set up properly then a rapid response can be expected, leading to good tenant satisfaction.

The major challenge in using this type of contract lies in establishing the correct lump sum each year. Basing the figure on recorded data will only prove an adequate approach if there is a large enough sample on which to produce a statistically sound figure. In addition to this, as with the use of historical costs for budgeting, real maintenance needs may be underestimated.

The operation of this approach requires high levels of goodwill on all sides, and is facilitated if extremely thorough, accurately costed condition data exists.

Term contracts

Under this type of contract, the Contractor is given the opportunity to carry out all work for the client, usually the subject of a works order, for a specified period of time, within an agreed geographical area. The work done is then normally priced in one of two ways:

❏ through the medium of an agreed Schedule of Rates, when it is termed a 'measured term contract'
❏ on a cost reimbursement basis, when it is normally called a 'Daywork Term Contract'.

In some cases, for larger items, a lump sum contract may be negotiated.

In essence, the term contract can be viewed as a specific case of other forms of contract described earlier. They have become increasingly used, in the measured term format, for maintenance work, most notably by the now defunct PSA.[1] PSA figures report that 40% of their minor works expenditure in the years from 1982 to 1989 was through measured term contracts, the next most widely used method being lump sum contracts for works between £25 000 and £250 000 in value.

BMI suggested that the most common basis for such contracts is a priced schedule. The most popular schedules were those developed by the client themselves, followed by the PSA[2] *Schedule of Rates* and the *National Schedule of Rates*,[3] which accompanies the Joint Contracts Tribunal (JCT) standard form of measured term contract. The BMI concluded that, whilst terms varied from one to six years, three years is the general norm. For contracts over a year a fluctuation provision is given, based on either published cost indices or annual updating of the rates. The PSA developed a number of standard schedules, and the normal tendering process would make use of the percentage on or off approach described above.

The work of producing the *Schedule of Rates*, initiated by the PSA, is now undertaken by Carillion Services Ltd. It is published approximately every five years by The Stationery Office (TSO)[4,5]

The *National Schedule of Rates* was launched by the Society of Chief Quantity Surveyors and the Building Employer's Confederation in 1982, and it is now published annually by NSR Management Ltd.[6]

Measured term contracts have been found to be most beneficial where the client has an ongoing need for maintenance, or other minor new work. A prerequisite is that there is a large enough workload to offer sufficient continuity for economic operations. They are also of value in assuring a rapid response to work requests. To ensure that this happens, it will be normal to specify response periods, and this requires a good communication system and means of monitoring and controlling performance.

In theory, the benefits of an assured work programme should lead to keen pricing. However, this may not mean the lowest price for every rate, and hence the lowest price for every order. It must be viewed as a package, and judged in terms of its ability to give the lowest overall price.

Under normal operation, the percentage on or off should allow for fluctuations over the period of the contract. This process is rather uncertain, as wage rates and material prices will change at a different rate, and as their proportions in the workload are not readily predictable, the probability that such an adjustment is accurate is very low. Users of the PSA *Schedule of Rates* can make adjustments to the base rates by reference to DTI data.[7]

Perhaps the major benefit to be realised through the letting of an assured programme of work is the opportunity for the development of a good client/Contractor working relationship, with corresponding improvements to quality of service.

It is also claimed that the measured term contract offers potential savings in time and administrative costs, compared with the letting of a multiplicity of contracts. Whilst this may be true for the pre-contract stage, it is a doubtful claim when total costs over the life of the project are considered.

Difficulties may also be encountered at a changeover point, where the Contractor for the previous period is being replaced. There will be inevitable disruption to work programmes. The sanction, retained by the client, to terminate a contract under given circumstances, may also cause severe disruption, and can leave a large number of incomplete jobs. These factors may place pressure on the client to retain a Contractor under circumstances where he should properly be replaced.

Contractor selection

Depending on which contractual form is adopted, there will be a procedure for selecting a Contractor. In any type of project there are certain good practice rules that should be adopted, in order to safeguard the client's interests. The objective is to select from those available the Contractor who is most likely to satisfy the client's requirements. Price will only be one of these requirements, and the specific characteristics of maintenance work give rise to other performance parameters. These largely stem from the need to provide a service, rather than a specific product, and this distinguishes maintenance from other types of work. The ability of the Contractor to be able to respond promptly, and to efficiently execute a range of small tasks, often on a frequent basis but without the benefit of much forward planning, is of major importance.

Up until the 1980s few Contractors specialised in maintenance contract work. However, diminishing new-build workloads raised the interest of a wider range of companies, who previously would not have considered this type of work. Historically, maintenance contract work was attractive to small Contractors, who were considered to be the ones most appropriately structured. The interest of larger organisations in maintenance work is attributable less to the attractiveness of maintenance work than to difficulties in other sectors. Therefore, whilst the increased competition for work may present benefits on a cost basis, there is a need for caution.

The selection of a Contractor should be made on as wide a basis as possible, taking care to compare the Contractor's known or assumed abilities with those required. Within this context reference should also be made to developments in partnering and the outsourcing of facilities management packages discussed in Chapter 1.

The following are general factors that should always be taken into account in Contractor selection.

(1)   Reputation
Familiarity with a Contractor's capability for executing the type of work under consideration is always a valuable starting point. Volatile market conditions impose

the need to give careful consideration to this factor, particularly with new Contractors seeking to break into the maintenance market. It may well be prudent to question motives, as the Contractor may see this market as a temporary one to make use of spare resources. Evidence of real commitment is not easy to establish.

(2)   Resources
It is crucial that the Contractor possesses sufficient resources to execute the work, but in the case of maintenance work it is important to consider type and quality as well as quantity. Much can be learned about a Contractor's ability by studying his physical resources, as represented by his stores, offices and workshops, and about his labour resources from knowledge of previous contracts. Of particular importance is his management set-up, and careful scrutiny of its ability to handle the specific requirements of managing maintenance work is essential.

(3)   Workload and availability
Whilst this must always be considered when selecting a Contractor, it must be borne in mind that the nature of maintenance work may demand a service measured in terms of response times. This requires a rather different type of commitment from the Contractor than on new-build work, where he is able, within certain limits, to plan his labour commitment for a project.

(4)   Price
In many cases this is the sole criterion used for judging a Contractor, and this view is clearly a very narrow one. It must be remembered, however, that in many cases, particularly in the public sector, accountability is a prime requirement, and there is a need for the placing of contracts not only to be fair, but to be easily seen to be so. This can often be used as an excuse for an oversimplified approach. Developments in housing management by housing associations, where a range of performance criteria are employed, is encouraging in this respect.[8]

Selection procedures

A simple way of classifying procedures for selecting a Contractor is by the degree of competition.

(1)   Open tendering is a method whereby a contract is advertised, and all Contractors are free to tender, without any prior enquiry with respect to their ability. This type of tendering is now rarely used, even where accountability is a major consideration, mainly because of quality and performance concerns.
(2)   Selective tendering is a refinement of the above method, designed to remove its major drawbacks.[9] Tenders are invited from a list of Contractors, which has been compiled having regard to their reputation and, just as importantly, suitability for the type of work in question. Large client organisations may hold permanent lists, which are updated at intervals, but in some cases special lists are compiled for a particular project. In the latter case, an advertisement will invite Contractors to apply to go on a list.

(3) Negotiated contracts involve the client inviting a tender from a known Contractor, who is judged to have the necessary qualities for the task at hand. This tender may then be subject to further negotiation regarding conditions and prices. Whilst the level of competition, particularly in cost terms, is strictly limited, it can provide a very satisfactory outcome where other requirements, such as time and quality, are considered to be of significant importance.

Tendering can be carried out in a single stage, based on one set of contract documentation, usually at the end of the design stage. So-called traditional competitive tendering for a lump sum contract typically operates under this process. Measured term contracts, similarly, can be single stage.

In some cases, multi-stage tendering may be deemed appropriate. For example, in new-build work, the Contractor may be selected by means of a first stage competitive tender, based on documentation related to preliminary design information, which provides a level of pricing for subsequent negotiation. The second stage consists of accurate pricing of the completed design, the Contractor having been involved in design development following a first stage appointment.[10] This is a procedure to be followed when it is considered beneficial for the Contractor to be involved in the design process.

From the legal viewpoint, the tender represents the offer component of the contract. This offer may be made at any one of a number of points in time, and may be based solely on price or other factors. The offer may be in the form of a fixed price, either stated as a lump sum at the time of tendering, or to be arrived at on the basis of a Schedule of Rates after completion, or for an indeterminate sum of money arrived at through cost plus a percentage, or fee. The offer may relate to a single job, or to a series of jobs over a defined period of time.

Following the offer, the client or his representatives must make a selection, or 'adjudication', as it is commonly termed. This process may be simple, if price is the sole criterion, but if offers are to be judged on a range of issues then it is important that the relative importance of these criteria is clearly defined. This may be done in an informal way, based solely on judgement and experience, or recourse may be made to a weighting system.[11] It is not essential for there to be a written contract, provided the requirements of common law are met. For example, small jobs are often carried out on the basis of a simple verbal or written exchange. In this case, the absence of written conditions requires the use of some ground rules provided by the following implied terms.

(1) The building owner must allow the Contractor to enter a building at the necessary time for the purpose of executing the work, give necessary instructions within a reasonable time, and not obstruct the Contractor in the execution of his work.
(2) The Contractor must do the work in a workmanlike manner, and complete it within a reasonable period of time.
(3) In addition, there is an implied warranty that any materials supplied are reasonably fit for the purposes for which they will be used, and of good quality.

Clearly there is a great deal of uncertainty in the interpretation of these terms, and in the case of dispute recourse would have to be through common law, with the attendant costs of such actions. For most contracts above a minimum size, the Contract Conditions should be in writing, and there is a range of standard forms of documentation that can be used for this purpose.

Contract documents

Contract documents are those 'which can be identified as containing the terms of a concluded contractual agreement between the parties which was made in, or reduced to, or recorded in writing'.[12] Documentation for building and construction contracts usually consists of:

❑   Articles of Agreement
❑   Conditions of Contract

and one or more of the following, depending on the type of contract:

❑   Drawings
❑   Specifications
❑   Bills of Quantities, either firm or approximate
❑   Schedules of Rates

The articles of agreement represent the core of any particular contract, and contain the definitions of the parties, a description of the scope of the Works, the price mechanism, and the persons to operate under it. It is here that the signatures of the parties and witnesses go. In most standard forms of building contract the articles can be obtained as a pre-printed form, usually along with the conditions, with spaces left for relevant infill.

The conditions of contract contain the detailed provisions as to how the contract is to be operated. For example, the rights and duties of the parties are more clearly defined, and detailed arrangements given for the treatment of variations to the contract, methods of payment, and so on.

A major component of the contract will be price, and the Contract Conditions must provide proper safeguards for all parties. The concern of the building owner will be that they pay a fair price in relation to the quantity and quality of service provided. Payment mechanisms must, therefore, be carefully laid out to help ensure that this happens, and it is also useful if they assist the operation of a proper cost control system. The various standard contract forms have ways of dealing with this, often making using of a contract particulars section containing project-specific information.

In general there will be conditions relating to each of the following:

(1)   A statement of the lump sum, or the formula to be used in establishing the final cost. In the case of a lump sum, there should be an itemised breakdown of the work, with each item priced to assist cost control, and to form the basis for

evaluating any variations to the contract. At the limit this is a bill of quantities. Such an itemised breakdown is unlikely to exist for maintenance work.

(2)   A statement of the documentation that must be provided by the Contractor to support an application for payment. In a Daywork Term Contract, for example, this will probably include time sheets and material invoices.

(3)   The periods at which the Contractor will be entitled to receive payment for work completed. For lump sum contracts this will be based on an interim valuation of the work. For measured term contracts it may be on a monthly basis, determined by the value of orders executed.

(4)   Procedures should be laid down for varying the work, in terms of either quality or scope and, particularly in lump sum contracts, a proper procedure for ascertaining the value of variations. For schedule- or daywork-based contracts, payment should theoretically be received automatically. Problems occur with schedules where an item is not covered, and the contract should make provision for handling this. From a legal point of view, the major problem that occurs is when the scope of the work is altered to such an extent as to bring into question the continued validity of the contract.

(5)   In reimbursement contracts, with a fixed or percentage fee, it is necessary to provide a statement as to which costs are prime and which are part of the fee, i.e. deemed to be included in the add-on.

(6)   Reimbursement contract forms will commonly differ from other forms, especially lump sum ones, in the scope they allow the client to influence working methods and practices.

Standard forms that are available generally divide into those with and those without quantities. The JCT produce a portfolio of standard building contracts, which are widely used for a range of different-sized projects. In 2005 the JCT began a comprehensive revision of its whole suite of forms in a new-look format to improve their organisation and layout to make them easier to use.

The Standard Building Contract (SBC) is available with full quantities,[13] approximate quantities[14] or without quantities,[15] and is widely used, primarily for new-build projects of significant size and complex nature.

The Intermediate Building Contract (IC)[16] is recommended for building works of simple content, where the use of the SBC would be unnecessarily complex. IC is widely used because of its relative simplicity, even for large-scale projects. A further version of IC[17] was introduced in 2005 suitable for work that, although of a simple nature, requires part of the work to be designed by the Contractor to meet the Employer's requirements.

There has been some debate concerning the proliferation of forms of contracts and, in particular, many doubts as to whether they are being used in an appropriate way. Selection of the right one for any project is fraught with difficulty, but there are a number of forms available that are generally appropriate for maintenance work.

Minor Works Building Contract

The Minor Works Building Contract (MW) is available from JCT[18] for use where a lump sum has been agreed, and where an Architect/Contract Adminstrator (CA) has been appointed. This form is not suitable for use where a bill of quantities has been prepared, for works of a complex nature, or where it is intended to govern work by named specialists. Despite these restrictions, there exist many situations where this form can be used for sizeable projects.

An alternative version is also available for minor building work to permit part of the work to be to the Contractor's design.[19]

Where the full extent of a maintenance contract can be clearly defined, for example under a planned maintenance programme, the MW form is quite suitable. The main restriction to its use is clearly the complexity of the work, but where it consists of straightforward traditional construction, the provisions provide a good basis for satisfactory administration of the work.

Its use does not preclude the production of a bill of quantities to assist in arriving at a lump sum, but means other than the bill will be required for the operation of the contract. In other words, such a bill will not be a contract document, and the Contractor should therefore treat it with caution. It is envisaged that a priced specification or schedule will be required to value variations.

Similarly, the Contractor is not precluded from sub-contracting elements of the work, nor is the Architect or CA prevented from specifying, in the relevant contract documents, that certain specialist operations or materials are to be used. Such sub-Contractors and suppliers would then operate on the same basis as the Contractor's own.

The time span under which this form can operate is also strictly limited, as it makes no provision for fluctuations.

The form starts with the Articles of Agreement followed by the Conditions.

The Articles of Agreement name the parties and identify the work to be carried out, with reference to the contract documents, which may include one or more of drawings, specification or work schedules. The parties to the contract are to sign the contract documents, and the Contractor is to price the specification or work schedules, or provide a schedule of rates. The contract sum is given in the articles, along with the name of the Architect/CA, and there is provision for the naming of a planning supervisor where CDM regulations apply.

The contract-specific information is contained in the Contract Particulars section of the Articles, which requires blanks to be filled in, or where a provision is optional it may be struck out if not required. This part of the standard form sets out commencement and completion dates; the rate for liquidated damages for late completion; and the Rectification Period during which the Contractor is obliged to make good defects (usually three months).

The Contract Conditions are presented in seven sections, as follows.

Section 1 provides a schedule of definitions and interpretations of words and phrases used in the contract.

Section 2 is concerned with commencement and completion, and contains provisions for extensions of time. These are simple in form and state that an extension of time may be granted '. . . for reasons beyond the control of the Contractor, including compliance with any instruction of the Architect/CA under this contract whose issue is not due to the default of the Contractor . . .'. This is a very broad statement, and it has been held by some commentators to be potentially troublesome unless its application is restricted in some way, perhaps by amending the contract with a list of acceptable reasons. However, having said this, any amendment to a standard contract form should only be done after careful consideration and taking legal advice.

It is also clear that there is an onus on the Contractor to notify the Architect/CA immediately it becomes apparent that the Works will not be completed on time. The Architect/CA must then respond by granting a reasonable extension of time, provided the reasons for the delay are valid in this respect.

The correction of inconsistencies is covered in clause 2.4, and where these involve a change they are treated as a variation.

Clause 2.6 covers the Contractor's obligation to pay any fees (including any rates or taxes) legally demanded and any of the Statutory Requirements (defined in clause 1.1). Such fees and charges are not reimbursable to the Contractor by the Employer, unless otherwise agreed (for example, VAT under clause 4.1).

Provisions relating to defects liability during the Rectification Period (the default period is three months, unless the Parties agree otherwise) are made in clause 2.10.

Section 3 is concerned with the control of the Works, and contains several interesting provisions. If the Contractor wishes to sublet any portion of the work, he must obtain the written consent of the Architect/CA to do so, although this must not unreasonably be withheld.

Of great importance is the granting of apparently unrestricted power to the Architect/CA to issue instructions with which the Contractor must comply, and these instructions may be not only in respect of additions and omissions, but also related to the order, or period in which the Works are to be carried out. However, it is debatable as to whether this right extends to directing the Contractor to execute the work in a particular way. There seems to be a fine dividing line between these two positions.

The power to alter the period of the work, which presumably means in either direction, is also given. This is quite separate from the extension of time provision and such changes '. . . shall be valued by the Architect/CA on a fair and reasonable basis using any relevant prices in the priced Specification/Work Schedules/Schedule of Rates . . .'.

Whilst there is no provision for PC sums, as nominations are expressly excluded, there is the facility for Provisional Sums and these are covered under clause 3.7.

Section 4 deals with the important question of payment. Provision is made in 4.3 for monthly progress payments to be made to the Contractor, of the percentage stated in the Contract Particulars of the *total value* of the work (the default percentage is 97%, unless the parties agree otherwise, i.e. 3% is retained by the Employer). Within 14 days after the date of practical completion the Employer must issue a penultimate certificate, of the percentage stated in the Contract Particulars, of the *total amount* of the work (the default percentage is 98.5%, unless the parties agree otherwise) as far as it is ascertainable at the date of practical completion.

The Final Certificate clause 4.8 places an obligation on the Contractor to produce the documentation, reasonably considered necessary to determine the final amount, within the period stated in the Contract Particulars (the default position is three months, unless otherwise agreed by the parties), and specifies that the Final Certificate is to be issued within 28 days of its receipt, or of the issue of the Certificate of Making Good Defects, whichever is the later.

The contract provides an option for fluctuations in respect of contributions, levies and taxation to operate, as a default position, unless this option is deleted in the Contract Particulars. Where fluctuations are to operate they are dealt with as described in Schedule 2.

Section 5 deals with injury, damage and insurance, and the main things to note are the provisions relating to existing buildings. It is also worth noting that the amount of insurance cover the Contractor is required to take out should be related to the proximity, nature and use of adjoining premises, and presumably this should also include condition. The size of the job does not necessarily relate to the potential size of claim for which the parties might be liable.

Section 7 covers termination, giving the Employer the right to dismiss the Contractor if he '. . . without reasonable cause fails to proceed diligently with the work . . .' or if the Contractor becomes insolvent.

The most contentious of the two reasons for termination is the former. The Employer must clearly make sure that he is on firm ground, otherwise he lies open to an action from the Contractor and subsequent damages. It should be noted that the clauses refer to the termination of employment of the Contractor, and the contract remains in force.

The Contractor is given the right to terminate his own employment if the Employer defaults in a number of ways, such as failing to make progress or becoming insolvent.

Section 7 sets out the processes that may be used to settle disputes. Clause 7.1 suggests that the parties should seek to resolve disputes by mediation. However, this is only a recommendation rather than an obligation. If mediation is not used or fails to provide a solution, adjudication or arbitration processes can be used (in the case of the latter the Contract Particulars must state that the provisions of Article 7 and Schedule 1 are to apply).

JCT Standard Form of Measured Term Contract

JCT have produced a Standard Form of Measured Term Contract,[20] which includes Articles of Agreement, Conditions and an Appendix. At the time of writing a revised version of this form of contract (MTC 05) is expected later in 2006.[21]

The tendering Contractor offers to carry out, and complete, all orders for work, in accordance with the information set out in the Appendix, and with the Contract

Conditions of the MTC 90, at the rates listed in the Schedule of Rates. The latter is subject to the addition/deduction of what is termed the percentage A, which the Contractor enters in the appropriate space in Item 6 of the Appendix. This represents the Contractor's basic offer.

The tendering Contractor is then asked to insert percentage B, which is the percentage he requires to be added to works carried out as daywork under the contract. A percentage is to be entered against each of the following items under this heading set out in Item 9 of the Appendix:

- ❏ overheads and profit on labour
- ❏ overheads and profit on materials
- ❏ overheads and materials on plant
- ❏ overheads and profit on sub-Contractors.

A percentage C entry is then required for overheads and profit on non-productive overtime.

Articles of Agreement

The parties to the contract are identified in the Articles and the 'Contract Area' defined in the 1st Recital. The 1st Recital requires maintenance and minor works to be carried out in the Contract Area in accordance with the details set out in the Appendix, Items 1 to 5. These details include the minimum value of any one order to be issued under the Contract, the approximate anticipated value of work to be carried out and the Contract Period.

The 2nd Recital states that the Contractor has offered to carry out the Works based upon the terms of payment in the Appendix, Items 6 to 10.

Article 1 states that the Contractor will carry out the maintenance and minor works as set out in the Appendix, Items 1 to 5, and Article 2 states that the Employer will pay the Contractor in accordance with the terms set out in the Appendix, Items 6 to 10, and subject to the Conditions.

Article 3 identifies the Employer and Articles 4 and 5 identify the Planning Supervisor and Principal Contractor respectively as required by the CDM Regulations.

Articles 6, 7A and 7B identify methods of settling disputes or differences by adjudication, arbitration and legal proceedings respectively. If speed of resolution of a dispute is important for the Employer, adjudication would be the most suitable option.

The Appendix

The Appendix to the standard form is divided into several Items, and sets out a number of essential parameters for the administration of the contract.

Item 1 requires the entering of a list of the properties, in the so-called Contract Area, in respect of which orders can be issued under the contract. It also requires a description of the type of work for which orders may be issued.

The Contractor's right of access to the site, which may be an individual property, is covered under Contract Condition clauses 3.4 and 3.5. In essence, the Contract Conditions state that it is the responsibility of the CA to arrange access, unless it is stipulated otherwise in the preliminaries to the Schedule of Rates. There has been some criticism of this condition, on the grounds that it is normally more realistic for the Contractor to arrange his own access, and that this should be the norm contractually – that is, the standard form has it the wrong way round.

Item 2 specifies the maximum and minimum values of orders that may be issued under the contract. In some variations on this standard form, the tender invitation provides value bands, and gives the Contractor the option of putting a different percentage on or off (percentage A) against each band. It has been found in practice, however, that given this option, contractors still have a tendency to use a single percentage.

Item 3 gives the approximate anticipated value of the work to be covered by the contract. It is important to note, however, that under Contract Conditions clause 1.13 the Employer accepts no responsibility as to the actual amount of work that will be ordered, and that no change in the percentages A, B and C will be considered by the Employer if the actual value of the work ordered differs.

Item 4 specifies the Contract Period and its commencement date. It is suggested that this should never be less than a year, nor longer than three years. Contract Condition clause 1.15 requires that the Contractor must provide the CA with a programme for the Works, if so requested.

Under clause 2.1 of the Contract Conditions, it is a requirement that orders shall be reasonably capable of being carried out within the Contract Period, unless otherwise agreed. Clause 2.2 asserts that orders must be executed in accordance with any priority coding specified. The details of such a priority code must also be given in Item 4 of the Appendix. Each order that is given will require completion by a certain date, and there must be a mechanism for agreeing that an order is complete. The Contractor notifies the CA under clause 4.12 when it is considered that this is the case, and this date stands unless the latter disagrees within 14 days.

Clause 2.3 also provides for delays, and requires the Contractor to notify the CA, when it becomes apparent, of any matter likely to cause completion of an order to exceed the specified completion date. The clause refers to matters beyond the Contractor's control, and provided the CA is satisfied, the Contractor may then fix a new completion date accordingly. Note that if this goes beyond the contract completion date, the Contractor must still discharge his duties under the contract. This is referred to as 'fixing later date for completion', and is not an extension of time to the contract.

The Schedule of Rates to be used for the contract is referred to in *Item 5*, but actually specified in *Item 6*. The default position is to use the rates listed in the *National Schedule of Rates*, but the Employer is free to delete this reference and specify whatever schedule of rates he chooses, either a standard schedule such as the PSA *Schedule of Rates* or a bespoke one of his own devising.

Items 7 and 8 represent alternatives, depending on whether or not a provision is to be made for fluctuations. In the event of no fluctuations, *Item 7* is deleted. If fluctuations are to be provided for then *Item 8* is deleted and *Item 7* specifies the basis on which rates may be varied. In the case of the *National Schedule of Rates* being used, it is updated on 1 August each year, and any orders issued after that date may be valued using the new rates. If a different schedule of rates is used, then an appropriate formula will need to be entered.

Item 9 provides for entry of the percentages B and C, in respect of daywork and overtime working, and the appropriate labour rate. Clauses 4.4 and 4.5 lay down the procedures for the measurement of daywork.

> it shall be calculated in accordance with the definition of *Prime Cost of Building Works of a Jobbing or Maintenance Character*, subject to adjustment by percentage 'B'.[22]

The Contractor is required to give the CA reasonable notice of the commencement of any work, or material supply, which he considers should be paid for on a daywork basis, and must then submit returns, in a form required by the CA, within seven days of the end of the week to which the work relates. The standard form specifically notes several instances where daywork payment is appropriate.

❑ Under clause 4.3, where it is not practicable, or would not be fair and reasonable, to value work using the Schedule of Rates, or to deduce appropriate rates, then payment on a daywork basis is justified.
❑ Where the Contractor's access to execute the work is disrupted, then he may, under clause 3.4.2, recover the cost of unproductive time on a daywork basis. Similarly clause 4.7 deals generally with disruption to the work, and payment by daywork is specifically referred to.
❑ When an order has been cancelled, then the Contractor may also recover abortive costs through daywork, under clause 3.8.2.

Additional payment for non-productive overtime, and the addition to labour rates of percentage C is permitted under clause 4.6, which lays down the rules for this type of payment, and specifies the Contractor's responsibilities in this respect.

Clauses 4.8 and 4.9 of the Contract Conditions define the responsibility for measurement of the work. In the interests of simplifying administrative procedures, it states that work under a certain value can be measured by the Contractor. Above this value, the CA is responsible for valuation of the work.

The appropriate value at which this occurs is entered in *Item 10* of the Appendix. It is noted in clause 4.10 that this may be deleted, in which case the Contractor effectively becomes responsible for measurement of all the work. The detailed rules with respect to payment and dissent for either of these scenarios are covered by clauses 4.13 and 4.14.

Clause 4.11.1 allows for progress payments to be made on orders over a certain value. This amount is also entered in *Item 10*. For orders that fall below this level, clause 4.12 provides for payment to the Contractor within 14 days of their notifying the CA that the order is complete. This is subject to the more detailed conditions on responsibility for measurement of the work.

Variation or modification of the design, quality or quantity of the work, or supply, that the order comprises is subject to the rules laid down in clause 3.6.

Under clause 3.8 the CA may cancel an order, and this clause also provides for reimbursement of the Contractor for the costs he may already have incurred.

Items 11, 12 and 13 deal with the Contractor's safety policy, statutory tax deduction and insurance respectively. The last of these requires the entry of an amount of cover, percentage addition for professional fees and the annual renewal date.

Clause 8.1 states that, notwithstanding the duration of the Contract Period, the employment of the Contractor may be determined by either party, not earlier than six months from the commencement of the Contract Period, provided that a notice of 13 weeks, or some other period specified in *Item 14* of the Appendix, is given.

This so-called break clause is considered to be very important for this type of contract. The smooth operation of measured term contracts requires that the parties develop a good working arrangement. It was felt, when the standard form was being drafted, that this 'escape' clause was necessary to allow dissolution of the agreement if it became apparent that a good working relationship did not exist.

Item 15 permits the entry of a named Arbitrator to deal with disputes arising under the contract.

Other important features

There are several other aspects of the standard form that are worthy of further comment.

(1) The Contract Conditions do not contain any preliminaries or specification. These are normally incorporated into the Schedule of Rates, although they may form a separate document.

(2) The contract form does not allow for nominated sub-Contractors or suppliers. Clause 3.2 requires the Contractor to obtain the express approval of the CA prior to sub-contracting any work. In normal parlance, 'consent shall not be unreasonably withheld or delayed'.

(3) The contract gives the right for the Employer to supply materials or plant for any work under Clause 1.6, and the subsequent clauses lay down the ground rules for the ownership and responsibility for storage, and general well-being of any such item. This permits large organisations to keep a stock of standard items, and presumably take advantage of bulk purchasing of heavily used items. This may be on economic grounds, perhaps to enhance public accountability, or to guarantee the availability at all times of critical components.

(4) Clause 1.2.2 gives the Employer the specific right to place orders for similar work within the Contract Area, and within the contract time, with any other organisation, including his own workforce. This is probably a sensible safeguard in the eventuality of a contracted organisation being overcommitted, perhaps as a result of exceptional circumstances.

(5)   All orders have to be in writing, which may include drawings. No oral instructions can be given, although there is the provision for orders to be confirmed in writing following a verbal order.

(6)   There are no specific provisions for liquidated damages and, at the time of writing (2005), whether an Employer may recover at common law losses due to late completion has yet to be tested.

(7)   There is a six months' defects liability period, but no retention money under the standard form. It does, however, permit the Employer to recover from future payment the cost of carrying out remedial work not done by the Contractor.

The question of insurances is sufficiently important to merit special attention, and the Contract Conditions under this standard form provide a good example of the general requirements in this respect. The matter is dealt with under *section 6* of the Contract Conditions of the standard form, and can be considered under several headings.

(1)   Injury to persons and property and indemnity to Employer
The Contractor is required to indemnify the Employer against liability etc., arising under statute or common law, as a result of personal injury or death, caused by the carrying out of the order, except where this is specifically attributable to the actions of the Employer, or persons for whom he is responsible. Similarly, the Contractor has to indemnify the Employer against losses to persons or property arising as a result of negligence, breach of statutory duty, omission or default of any of his employees, or any other person on the site for the purpose of executing the order, with the exception of the Employer or his representatives. The reference to property does not include the work comprised in the order.

These provisions summarise clauses 6.1, 6.2 and 6.3. Clause 6.4 goes on to say that, without prejudice to these responsibilities, the Contractor must take out an insurance policy to comply with the Employer's Liability (Compulsory Insurance) Act 1969. It is required that the insurance cover must not be less than the amount inserted in *Item 13* of the Appendix.

The Contractor is required, under the contract, to provide documentary evidence that insurance has been obtained. If this is not done, the Employer may take out the relevant policies and recover the cost of doing so from the Contractor.

The effect of ionising radiation, or contamination by radioactivity from any nuclear fuel or waste, is specifically excluded under clause 6.4.5.

(2)   Insurance of existing structures
This is clearly of major importance in maintenance work, and under this standard form clause 6.5 requires that the Employer should insure the existing structures in respect of which orders are made, together with the contents that they own, or for which they are responsible, against loss or damage as a result of fire, lightning, explosion, storm, tempest, flood, bursting or overflowing of water tanks, apparatus or pipes, earthquake, aircraft and other aerial devices or articles dropped from them, riot and civil commotion. The exclusions under clause 6.4.5 are also relevant under this heading.

Provision exists for joint insurance by Employer and Contractor, and rules are given for the apportionment of costs.

Except when a local authority, the Employer is required to provide the Contractor with documentary evidence of the relevant policies as requested.

(3)   All risks insurance
The Contractor will normally be required to take out an all risks insurance policy, which provides cover against any physical loss or damage to work executed, or materials supplied, with respect to any order made, and any site materials. Under the JCT Standard Measured Term contract, the latter may include any materials supplied by the Employer.

Under clause 6.8 in the standard form, this specifically excludes:

❏   property that is defective as a result of:
   ○   wear and tear
   ○   obsolescence
   ○   deterioration, rust or mildew
❏   loss or damage due to defects in design, specification, material, or workmanship
❏   loss or damage due to war, invasion etc., radiation etc., as outlined in 6.4.5, and any unlawful act executed by an unlawful association, i.e. terrorism.

The clauses 6.9 to 6.13 give detailed requirements for all risk policies, and the ways in which claims should be handled for a variety of circumstances.

Jobbing Agreement

This is a comparatively new form of contract, introduced by JCT in April 1990.[23] It has a set of very simple Contract Conditions, covering about three sides of A4. It is capable of being used with the JCT standard form of tender, or with an organisation's own form. For example, a local authority's or housing association's standard works order is likely to be more appropriate than the rather lengthy JCT standard form of tender invitation.

Use of the JCT tender invitation has to be accompanied by a full description of the work required, with a specification and/or drawings. For many of the jobs for which this form of contract is intended, such an approach may be inappropriate. The following information is required to be entered on the tender form:

❏   start and finish date
❏   minimum amount of public liability insurance
❏   defects liability period
❏   name of the Employer's Representative.

It is also possible to place a verbal order by telephone, provided this is followed by a written order, the major stipulation being that fair and reasonable time periods are allowed, and that the price is reasonable. This scenario presupposes that the Employer has a list of approved Contractors, who are already in possession of a copy of the contract and have previously agreed fundamental issues, such as the work area and insurance coverage.

Contract documentation

The object of Contract Conditions should be to provide a fair and equitable legal framework, which will ensure that the work is carried out in a proper manner, and that the Contractor will receive reasonable payment. It is not intended to deal in detail with the legal relationships between the parties, but merely to tackle those features that have a particular relevance in the context of the work being executed.

It is, however, an essential requirement that the parties should have a clear understanding of their responsibilities and obligations, and there is therefore a need for supplementary documentation, which may take a number of forms, depending on the type of contract considered to be most appropriate.

The aims of this documentation can reasonably be summarised as follows.

(1) To provide a basis for the submission of a tender, and later evaluation of the work for payment purposes.
(2) To provide a clear picture of the job content or conditions, in a format useful for project management.
(3) The third objective, which is rather all embracing, is to provide a clearly defined supplement to the ground rules laid down by the Contract Conditions, for the execution of the project.

Contract documentation has to be clear and unambiguous, within the contextual limits that exist and, most importantly, in a form that is appropriate for the nature of the work and the conditions under which it is to be undertaken.

For much new-build work, it is assumed that the full scope of the works is defined through drawings, specification and bills of quantities, prepared prior to obtaining tenders, leading to a lump sum contract.

For maintenance work, it is rarely the case that sufficient information is available to describe the full scope of the work in precise detail prior to tender. Documentation methods may, therefore, vary from a detailed schedule of items to be priced to a broad statement of the service to be provided, accompanied by a statement of the end result to be achieved. Schedules of rates are therefore of great importance for contract maintenance work, and maintenance specifications will have particular characteristics.

Schedules of Rates

The Schedule of Rates is designed to have primary functions similar to those of a bill of quantities. In the first instance, it provides a means of comparing the tenders from a series of Contractors on the basis of price. To fulfil this function easily, some simplicity is desirable. A comparison is easy, for example, if Contractors are required to add a single percentage only. If, however, banded order values are given, with the Contractor able to put different percentages against each, then comparison is only easy if the relative proportion of work in each band can be readily predicted. In other variants of schedules, items may be grouped in some way, with the possibility of a different percentage for each group. Again, this may lead to problems of comparison when the tenders are being assessed.

In the case of an unpriced schedule, where the Contractor enters his own rates, there will be great uncertainty in assessing tenders unless the relative quantities of every item are known within reasonable limits. For these reasons, schedules for maintenance work are generally of the pre-priced variety.

Schedules may be produced on an ad hoc basis, following a similar pattern to a bill of quantities, for a particular job or, as is more usual, produced in a standard form. They are normally prepared to cover a range of repetitive jobs, and may be devised to meet the particular needs of an organisation, or may be standard ones that are more broadly based for general application.

The second function of a schedule is to provide a basis for payment, in conjunction with the Contract Conditions, and it is thus fundamental to the use of the measured term contract.

The decision as to whether to use an ad hoc or a standard schedule depends on a great many factors. Standard schedules will generally have the following advantages:

❑ They provide a relatively cheap control document for larger contracts.
❑ Because they are readily available, administrative costs in preparing contract documentation are reduced.
❑ They can be an integral part of an established set of procedures that have been well tried and tested, and fit in well with computerised systems.
❑ Both Employer's representatives and Contractors will be familiar with their use, which reduces the perceived risk of all the parties, and promotes good working relationships.

On the other hand, they do have some disadvantages:

❑ Their comprehensiveness can make them unwieldy, particularly for smaller contracts.
❑ The standard form may not be appropriate to the needs of the particular estate.
❑ They need to be able to integrate properly with established management systems.

Bespoke schedules may be more relevant in some cases as they have the advantage of conciseness, and can be tailored to suit the specific needs of a contract or of a particular client. On the other hand, there may be a lack of familiarity with their use, and this may cause operating difficulties and increase costs. They may also lead to increased risk levels, and lead to a cautious pricing policy by Contractors.

The old PSA of the DoE produced a series of standard schedules for their own extensive range of work, and these were used mainly with term contracts. These types of schedules provide a basis for estimating costs, tendering and valuation during execution. The PSA schedules were used for new build, as well as maintenance and repair, and published monthly percentage update indices assisted in the administration of a fluctuation clause.

The *National Schedule of Rates* comprises a suite of schedules to cover general building maintenance work and maintenance work for engineering and electrical services. Separate schedules, using items selected from the Building Schedule, are available for painting and decorating, and housing maintenance.[24,25,26,27] The Building Schedule contains approximately 10 000 priced items of repetitive general maintenance work. The format is much like a bill of quantities, but includes separate material, labour and plant costs. The *National Schedule of Rates* is re-issued in updated form each year.

A properly maintained Schedule of Rates provides other important subsidiary functions.

(1)   The priced schedule enables maintenance managers to estimate, with some accuracy, the cost of a maintenance programme, which assists in the evaluation of alternative strategies and the framing of bids for funding.
(2)   A properly maintained schedule will enable the manager to value each order, and thus provide data for an efficient cost control system. This will further enable them to control the flow of work, and perhaps evaluate alternative priorities on an ongoing basis.
(3)   Some types of schedule can be adopted to form the basis of an ordering system, and may considerably simplify administration. The manager may effectively be purchasing a maintenance service, rather like purchasing an item of goods. Theoretically, the order value and invoice can be the same.

The BMCIS originally produced a price book, intended as an aid to estimating maintenance work. This price book has also been used as a basis for measured term contracts. Its use in this way is extremely limited, but it does have the merit of simplicity, under some circumstances.

The contents of a Schedule of Rates will vary in detail, but essentially should include the following sections.

(1)   A set of preliminaries, which define the scope of the work and fully describe the conditions under which it has to be executed. This will include reference to Contract Conditions; for example, it may refer to the standard form to be used and any exclusions or modifications that will be made to it. In essence, like the preliminaries section to a bill of quantities, it should bring to the contractor's attention the major issues that will affect the pricing of the work. A set of standard preliminary clauses will normally be used, but these must reflect the conditions for that specific contract.
(2)   If a standard form of contract is being used, such as the JCT Measured Term, then this and its appendices will cover the major items, so that the preliminaries need only be a simple list.

It is necessary to include sufficient specification information to enable materials and workmanship to be properly defined and controlled, although these should be as simple as possible. The larger organisations will carry a set of standard specifications, and there is also a National Building Specification for maintenance work.

(3) The actual Schedule of Rates can take many forms. The essential requirement of all of them, however, is that they should be relevant to the work in hand.

The normally accepted practice is for the schedule to contain a full description of the item, including labour and materials, with a price attached to it. In the past schedules have been produced where the price is only for labour, with material costs being reimbursed at cost. Schedules of this type will not effectively perform the important subsidiary management functions, unless separate material costs can be easily added in. Clearly, this makes cost control and budgeting more tedious.

It has also been suggested, notably by the Audit Commission, that schedules should include some information on the predicted relative occurrence of each item.[28] Whilst this is, in theory, useful for a contractor, it would seem that the benefit can only be fully realised if the Contractor is given the option of using a variety of percentage add-ons. This then destroys the simplicity of approach that is generally endorsed by the majority of building maintenance professionals.

An ideal schedule would contain the exact description of the work to be carried out on any order that might be issued under the contract. The range of combinations possible, however, is so large that this is not feasible. In practice, a reasonable balance has to be sought, and this requires the use of schedules that are relevant in content, without containing too great a proportion of superfluous items. An item that is rarely used is, on balance, better left out, as the inclusion of unnecessary items can be misleading to the Contractor, as well as leading to a cumbersome document.

On the other hand, if relevant items are excluded, then a greater proportion of work will need to be valued by some other mechanism, including daywork. It will also reduce the efficiency of the schedule as a cost management tool.

A balance has to be struck between conciseness and completeness, and there is no panacea for this dilemma. The most important prerequisite for effective schedule usage is a good knowledge of the estate that is being managed, and past experience. Two types of schedule structure have been proposed to attempt to tackle this problem.

(1) The use of a very detailed schedule, which contains individual items of work for separate tasks, so that a job can be priced by building up overall rates from individual items of work, rather like a conventional estimating price book.

(2) A composite schedule containing items representing a range of possible descriptions for an item of work.

If, for example, it is required to replace a length of guttering, the detailed schedule would include a series of tasks, from which the price for replacement of guttering could be derived. On the other hand, a composite schedule would contain one or more items for replacing a length of guttering, including associated ancillary work. More than one composite description may be included for this, to reflect the different conditions under which it may have to be carried out, for example, its height above ground level.

The former has the benefit of accuracy and flexibility, whilst the latter has the merit of simplicity, in terms of both pricing and valuation. If the maintenance work is to be carried out on an estate that is fairly homogeneous in terms of building type and usage, then one would tend to the use of composite schedules, as there is less uncertainty attached to operations. If these conditions are not met then the use of detailed schedules may be preferred, in order to avoid excessive use of daywork but at the expense of increased administration costs. The size of orders will also be influential. A tendency to large orders provides a better justification for detailed schedules.

Specifications

The presentation of information in a specification can take a wide variety of forms, depending on its use and whether or not it is to be a contract document. For large new works the Architect may prepare a specification as part of the original brief to the quantity surveyor to assist in the preparation of the bill of quantities. For smaller new works, such as extensions and alterations, in the absence of a bill of quantities, the Contractor may be required to base the tender on drawings and a specification. For a large proportion of maintenance work there will be no drawings, so that the Specification takes on a great deal of significance. It is likely also that it is produced by a client's representatives, rather than by an architect or quantity surveyor.

For maintenance work a number of other particular difficulties should be noted, which will affect the form and content of the Specification.

For both planned and preventive maintenance, it can be assumed that some sort of scheme has been prepared and that the work can be reasonably well specified, thus permitting the use of standard specification clauses. Even on planned work, though, the uncertainty element associated with the nature of repairs may invalidate specification items.

For response maintenance, and repair work that is not predictive, a range of degrees of urgency may be associated with the work, involving the use of a priority coding system. The requirements in this respect may have to be included as part of the Specification.

In general, the following points need to be considered when designing specifications for maintenance work:

(1) Much maintenance work may be of the same nature as that carried out for new-build work, in which case standard specification clauses from a suitable source may be used.

(2) Under a planned or preventive programme, there will be a great proportion of the work where the extent is clearly defined, and for which standard maintenance specification clauses exist either in house or nationally. However, the execution of an item may bring to light the need for additional repair work that is a prerequisite for the completion of the planned item. Many specialised maintenance clauses are written in such a way as to provide sufficient flexibility to allow for this work to be carried out at the same time; for example, a standard clause for

the inspection of rainwater goods may also include an instruction for carrying out repair work. Similarly, internal decoration specifications will normally allow for repairing defective plasterwork when this is identified as part of the operation.

(3)   There are also operations that fall broadly under the heading of repair work, which in themselves are repetitive and will, therefore, be the subject of a set of standard specification clauses. Examples of this might include underpinning, or injected damp-proofing.

(4)   Much maintenance work is of an unknown quantity or type. Defects often manifest themselves in a number of ways, without the real cause being apparent. Specification clauses will therefore need to consider the work necessary to:
❑   determine the cause of the defect;
❑   eradicate the effects of a defect;
❑   rectify the defect.

The timescale for each of these operations may be different in terms of urgency. Serious water penetration due to a flat roof failure may require emergency action to clear up initial damage and to effect a temporary repair, followed by investigation and action to provide a permanent solution.

Whatever its content, it is essential that the Specification is clear and unambiguous, and a great deal of care has to be taken to ensure that items that affect the execution of the work are properly and accurately described. For example, where soil conditions are important, then the Specification must properly consider them. There are clear legal implications resulting from the production of an inaccurate document.

Assistance in drafting individual clauses is available as a computer-based subscription service from NBS, the publisher of the *National Building Specification*.[29] NBS, a part of RIBA Enterprises Ltd, provides a library of pre-written technical specification clauses and preliminaries for selection and editing to produce a project specification. The software package facilitates the writing of specifications on screen, referring to other technical information such as British Standards.

The PSA also produces a Specification for Minor Works,[30] and the Construction Project Information Committee[31] have produced useful supplementary advice.

In terms of structure, a specification document is normally divided into three main parts:

(1)   Preliminaries
It is often an accepted practice to include the preliminaries as part of the Schedule of Rates. However, wherever they are included in the contract documentation, the preliminaries give a set of provisions that govern the general conduct of the contract, and the extent of the Contractor's liabilities and obligations. These must be prepared very carefully, as they have to meet the specific requirements of the contract in hand. Generally, this section will include information on each of the following matters:

❑   a general description of the work
❑   the form of contract to be used, together with details of any modifications to it

❏   the contractor's obligations with respect to the provision of scaffolding, plant, etc.
❏   the requirements for the provision of huts, both for storage and for the administration of the contract, and to meet the welfare requirements of the site personnel
❏   an office for the employer's representative, if there is need for one
❏   the need to provide water, lighting and power for the execution of the work
❏   protective measures to be taken, especially where the work has to take place in occupied buildings
❏   information regarding times of access to the buildings, restrictions on working methods and other items that will affect the method and progress of the work.

(2)   Materials and workmanship

This section describes the quality of the materials to be used, methods of construction and standards of workmanship. Specifications for new works normally follow the order in which the work sections are given in the *Standard Method of Measurement of Building Works*,[32] or the CPIC *Common Arrangement*.

For many smaller projects, not all of these sections will be required, and there may also be merit in grouping some sections together. In many cases, it rapidly becomes apparent that a different approach is required. In particular, for alteration and repair work it is common practice for the ordering of the Specification to follow the sequence in which the work is to be executed on site.

Irrespective of the sequence in which the Specification is written, the clauses within each section are normally grouped as follows:

(1)   clauses of general applicability to the work section, which may include general job descriptions, and items such as plant, cleanliness and material storage
(2)   materials, and their preparation, which may be described in the following ways:
  ❏   by a full description, stating desirable and undesirable features, and any tests with which they should comply
  ❏   stating the relevant British Standard, which is considered to be a minimum quality, together with any Agrément Certification, and noting that in many cases more than one quality standard may be called for within a material, depending on its application
  ❏   specifying a proprietary brand, or naming a particular manufacturer or source of supply
  ❏   by giving a prime cost sum and an outline description of the material.

General adjectives such as 'best' or 'first class' should be avoided, unless they are recognised terms used to describe the particular quality being specified. The term 'other equal and approved' also needs to be treated with some caution.

There is a strong case for grouping materials in a separate section to avoid repeating the description wherever it is encountered, or to eliminate complex cross-referencing. The workmanship requirements would then follow the materials groups.

The workmanship clauses should state precisely the standards expected, giving details of any constraints on the method of working. Reference can also be made to relevant standards and codes of practice. These clauses will normally be produced on an operational basis: that is, following the sequence of site operations.

(3)   The Works
This section describes the Works to be carried out, using the materials and workmanship set out in the previous sections. Whereas the information for writing the Specification for new work comes from design drawings and schedules, much of the information concerning work for existing buildings must be gathered on site.

A draft specification is best written on site, and this enables each item to be fully considered and followed through. Taking rough notes, and writing descriptions later, takes longer and increases the possibilities of mistakes and errors. This underlines the importance of well-structured survey information.

It is helpful if the order of clauses in the work section is presented in such a way that each item is successively met in an ordered walk through the building, and there is consistency with survey methodology.

References

(1)   Building Maintenance Information (1993) *Measured Term Contracts.* Special Report. RICS, London.
(2)   Property Services Agency (1990) *PSA Schedule of Rates for Building Works.* PSA (DoE), London.
(3)   Society of Chief Quantity Surveyors in Local Government *National Schedule of Rates for Building Works.* Building Employers Confederation, London.
(4)   Carillion (2005) *PSA Schedule of Rates for Building Works*, 9th edn. The Stationery Office, London.
(5)   Carillion (2003) *PSA Schedule of Rates for Property Management: Composite items for maintenance, repair and improvement works*, 1st edn. The Stationery Office, London.
(6)   NSR Management (Published annually) *National Schedule of Rates for Building Works.* NSR Management, Aylesbury 2005.
(7)   DTI (2006) *Adjustments for Measured Term Contracts – Updating Percentages.* Tudorseed Construction, Hornchurch.
(8)   National Federation of Housing Associations (1987) *Standards for Housing Management.* National Federation of Housing Associations, London.
(9)   Joint Contracts Tribunal (1994) *Code of Procedure for Single Stage Selective Tendering.* JCT, London.
(10) Joint Contracts Tribunal (1994) *Code of Procedure for Two Stage Selective Tendering.* JCT, London.
(11) Janssens, D.E.L. (1991) *Design-Build Explained.* Macmillan, London.
(12) Powell-Smith, V. & Chappell, D. (1983) *Building Contract Dictionary.* Architectural Press, London.
(13) Joint Contracts Tribunal (2005) *Standard Building Contract with Quantities (SBC/Q).* Sweet & Maxwell, London.
(14) Joint Contracts Tribunal (2005) *Standard Building Contract with Approximate Quantities (SBC/AQ).* Sweet & Maxwell, London.
(15) Joint Contracts Tribunal (2005) *Standard Building Contract without Quantities (SBC/XQ).* Sweet & Maxwell, London.

(16) Joint Contracts Tribunal (2005) *Intermediate Building Contract (IC).* Sweet & Maxwell, London.
(17) Joint Contracts Tribunal (2005) *Intermediate Building Contract with Contractor's Design (ICD).* Sweet & Maxwell, London.
(18) Joint Contracts Tribunal (2005) *Minor Works Building Contract (MW).* Sweet & Maxwell, London.
(19) Joint Contracts Tribunal (2005) *Minor Works Building Contract with Contractor's Design (MWD).* Sweet & Maxwell, London.
(20) Joint Contracts Tribunal (1999) *Standard Form of Measured Term Contract (incorporating amendments 1: 1999, 2: 2000, 3: 2001, 4: 2003) (MTC 98).* JCT, London.
(21) Joint Contracts Tribunal (1975) *Measured Term Contract (MTC 05).* Sweet & Maxwell, London.
(22) Royal Institution of Chartered Surveyors (1975) *Definition of Prime Cost of Daywork carried out under a Building Contract.* RICS, London.
(23) Joint Contracts Tribunal (1990) *Jobbing Agreement.* JCT, London.
(24) NSR Management (2005) *National Schedule of Rates: Painting & Decorating.* NSR Management, Aylesbury.
(25) NSR Management (2005) *National Housing Maintenance Schedule.* NSR Management, Aylesbury.
(26) NSR Management (2005) *National Schedule of Rates: Mechanical Services in Buildings.* NSR Management, Aylesbury.
(27) NSR Management (2005) *National Schedule of Rates: Electrical Services in Buildings.* NSR Management, Aylesbury.
(28) Audit Commission for Local Authorities in England and Wales (1986) *Improving Council House Maintenance.* HMSO, London.
(29) www.thenbs.com
(30) Property Services Agency (1990) *PSA Minor Works Specification.* PSA, London.
(31) Allott, T. (ed) (1998) *Common Arrangement of Work Sections for Building Works,* 2nd edn. Construction Project Information Committee, London.
(32) RICS (1998) *Standard Method of Measurement of Building Works,* 7th edn. RICS, London.

Chapter 9
The Execution of Building Maintenance

Introduction

Of all the considerations relating to the execution of maintenance work, perhaps the strategic issue of who actually does the work is the most important. Traditionally, there have been significant differences of practice between public and private sectors in this respect, although this is changing somewhat as discussed in Chapter 1. It remains true, however, that many of the principles of maintenance management practice were established in the public sector.

Within this sector, there was a common presumption that maintenance work would largely be executed by a DLO, whilst in the private sector only work of a small jobbing nature was normally carried out by a limited in-house workforce. Traditionally, the degree of interest shown in contract maintenance work by private sector contractors tended to be somewhat limited.

The position at the time of writing is more fluid, and this is due to two major factors:

❏ the diminished role of DLOs and other in-house maintenance services
❏ the rapid increase in outsourcing particularly in the context of comprehensive facilities management packaging as discussed also in Chapter 1.

The debate with respect to who executes maintenance work in the public sector has been in the past an acrimonious one, with a tendency for the battle lines to be drawn along party political boundaries. Politically, these battle lines have become blurred and the issue is less sensitive, although still giving rise to some controversy in terms of quality of service delivery as a whole across the public sector.

Direct labour versus contracted-out maintenance

Direct labour for many years proved more popular for maintenance and repair work than for new-build work in the public sector. An Audit Commission[1] analysis in 1989 concluded that DLOs were market leaders in maintenance work, because of their size and approach to maintenance work and other work of a small jobbing nature.

This good performance was explained by increased efficiency of the organisations in the execution of their work or perhaps, to the surprise of many, as being due to the fact that they were not as inefficient as had been claimed by many of their critics.

The arguments for and against direct labour organisations

Arguments about the benefits of direct labour in local authorities tended to divide into:

❑   ideological
❑   debate concerned with financial, managerial and technical performance.

It is the latter of these that has now become the focus, and it is perceived performance that is the prime consideration in the political arena. The major argument has been that direct labour has greater flexibility for carrying out maintenance work, particularly emergency items, and by specialising in this type of work acquires the essential experience at both at management and operative level. Following on from this is an argument to do with the sensitivity of much of their operation, such as, for example, social housing maintenance. The very real political environment within which local authority maintenance is carried out requires particular methods of management and execution. Indeed, it was felt at one time that many private contractors might be reluctant to tender for work in such a context.

These arguments for many years carried much credibility in that there were comparatively few private contractors with the relevant experience. The growth of FM and increasing use of outsourcing has changed this somewhat. It is certainly true that there are now more contractors involved in maintenance operations, with firmly established experience.

Judgements may now be becoming more sophisticated. In assessing performance, cost is not the only factor to be considered, and quality of service is an essential requirement. This is true in all sectors, but within local authorities it has to be viewed in a social and political context. The promulgation of best value and the drive towards Public Private Partnerships discussed in Chapter 1 has, it could be argued, both freed up the public sector and at the same time encouraged a climate where competition is now not just about lowest cost.

The cost versus quality issues are, of course, intrinsically linked. There is a strong technical aspect to the quality argument, in that efficient maintenance increases life cycles, and this, in the long run, produces an overall saving in expenditure.

Another key factor is the relationship between planned and unplanned maintenance. For emergency repair work, direct labour is likely to have an edge in terms of pure quality measures, although there are problems associated with this. In particular is the significance of the definition of emergency work and what can be realistically put into a planned programme. Best value principles concentrate the mind somewhat in this respect.

These arguments, largely centred around public sector maintenance, identify many of the issues to be considered by any estate or building manager.

In many instances, in practice, both direct and contract labour will be used, and acceptance of the rationale for this is important for all decision makers. The real decision-making should especially be concerned with the most appropriate mode of execution, having regard to both context and corporate objectives.

For a public sector organisation, both context and objectives will be influenced by social and political factors as well as economic ones. In the private sector, commercial imperatives may be the dominant force. In both cases perceptions and notions of accountability should not be ignored, notwithstanding the differing scenarios within which each of these exists.

The decision-making process

A simple, if imperfect, division of maintenance work is into emergency or unpredictable work, and planned or predictive maintenance. It can be assumed that the latter lends itself more to the use of contracting. The criterion for choice will almost inevitably be price, although even with competitive tendering there will be some scope to select contractors to go on to a list on the basis of the quality of service they are able to provide, based perhaps on historical data.

Emergency or non-predictive maintenance presents more problems, in that although it may provide a powerful argument for maintaining a pool of direct labour, some of it may be contracted out.

The major debate has to be about the proportion of work to be allocated to each, which will normally be a question of deciding on the size of the direct labour pool (figure 9.1), and the setting up of appropriate contractual arrangements to provide the client body with a suitable degree of control and accountability. The latter will also be true for direct labour although protocols and the appropriateness of various performance measures may vary.

A problem that arises is that a contractor, whether private or an internal works department, in order to ensure its own survival, has to make commercial decisions. If it is to execute emergency repair work with the quality of service required by the client department it has to ensure that it has resources available to it, such as a labour pool. This may require a pool of labour in excess of that normally required by day-to-day operations, with the attendant costs this imposes. There is clearly a relationship between cost and service quality, and this is an essential feature of setting up any internal or external contracting arrangement. This identifies the key requirement for all outsourced maintenance or facilities management operations to have proper control and evaluation measures in place.

Issues of service levels and performance measurement outlined below should therefore be seen in this light and viewed within the changing context within which maintenance operates, as outlined in Chapter 1.

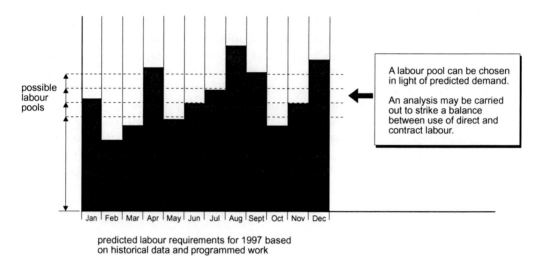

possible
labour
pools

A labour pool can be chosen
in light of predicted demand.

An analysis may be carried
out to strike a balance
between use of direct and
contract labour.

| Jan | Feb | Mar | Apr | May | Jun | Jul | Aug | Sept | Oct | Nov | Dec |

predicted labour requirements for 1997 based
on historical data and programmed work

Figure 9.1 Choosing an appropriate labour pool.

Service level agreements

A service level agreement is one of the most important vehicles for managing risk in
a contracted-out or outsourced service. The background given above and the discussion in Chapter 1 underline that such processes carry a number of risks:

❑ financial
❑ performance
❑ confidentiality/sensitivity/security
❑ parties going out of business.

In risk management the general principles are that the risk can be controlled,
accepted/paid for or transferred. Indeed one of the motivational forces for an organisation engaging in an outsourcing venture is to transfer the risk associated with carrying
out an operation in house. In other words the organisation may be transferring a liability, perhaps only for a short- to medium-term gain. Risk is also transferred every day
to insurance companies with the transferee paying for somebody to take the risk away.
One of the driving forces for the sale of local authority housing is to shift the liability
for maintenance out of the local authority on to the building owner. The new owner
will be accepting the risk associated with maintaining the property.

Another approach is to judge that the risk is worth taking and be prepared to pay
for any consequences that flow from it. In financial markets, for example, this may be
through imposing high rates of interest for venture capital.

In terms of controlling risk, it is normal for contractual arrangements to apportion
risk between the parties, and an outsourcing arrangement is in this sense no different.
However, as we have seen, the key factors are the fact that we are contracting out a

service – in this case, maintenance management – and may be entering into the arrangement as a long-term venture rather than for a project with a finite life. Such contracts will therefore have as an addendum to the contract a service level agreement (SLA) that defines the conditions of the services to be provided – for example, response times.

The key requirements for the SLA are that it should:

❏   be unambiguous
❏   be transparent – i.e. open to audit
❏   clearly describe the services to be provided
❏   specify a time period
❏   set out protocols through which the service is to be delivered – for example report-
      ing procedures, access arrangements
❏   define a methodology for payment
❏   detail a performance monitoring system and, by implication, give quality levels to
      be provided, in terms of clearly defined metrics, key performance indicators (KPIs)
      or benchmarks
❏   quantify rewards and penalties
❏   provide a clearly set-out set of procedures for dealing with dispute/problem
      resolution.
❏   give provisions for review and revision of the agreement.

Within these broad-brush requirements there is clearly a lot of detail. A search of the internet will reveal a great many sites providing guidelines and examples in terms of contents.

Performance measurement

The issue of performance measurement is really about controlling service provision, and in common with all control systems it implies that we need:

❏   a measure or metric
❏   measurement
❏   a yardstick against which to judge what is being measured
❏   evaluation
❏   a response.

In principle, if we wish to assess performance in any activity we can take an approach that is either criterion referenced or norm referenced. In criterion referencing, a required criterion or outcome that must be met is defined, perhaps in terms of levels of achievement: for example, in categories such as unsatisfactory, satisfactory, average, good or excellent. Care and skill is needed to devise very specific metrics in order to accurately categorise the level of performance achieved.

In norm referencing, we evaluate performance in terms of comparative measures against an expected norm. This assumes that we have data from a meaningful sample of results and can place the results from the service under evaluation into a rank within this sample. We mark or assess performance by where in 'the class' it falls. This approach leads logically on to the concept of benchmarking.

Benchmarking

Benchmarking is a valuable tool used for identifying and then implementing best practice in a process of continuous improvement.

The process involves addressing the questions:

❏ Who does better?
❏ Why are they better?
❏ How do we improve our performance?

Through benchmarking, an organisation measures performance in an activity and compares it with other operators, internal or external to the organisation, so as to norm-reference itself. The objective then set is to effect a series of measures to improve performance so as to achieve or exceed the same level as the best in the class. Some of the nuances are rather misunderstood as it is often criticised on the grounds that not everyone can achieve best in class. However, it should be understood that as a measure for driving up overall standards it can be very effective if organisations recognise this and embrace it.

Its use in the delivery of public services is widespread as it allows reliable comparison of services delivered in house with those contracted out, and by setting objectives in relation to these the belief is that the quality of public services will be driven up.

Four distinct types of benchmarking can be identified and they all have their relevance and potential for application to performance monitoring of FM services in general and maintenance in particular.

(1) Internal
This compares performance amongst units within an organisation by identifying common criteria. Examples could be individual retail units within a chain of shops, individual sites for a contractor or individual branches of a bank.

Research shows this to be an effective way of improving performance as each unit seeks to become the best in the organisation, leading to an overall levelling up.

(2) External or competitive
This compares performance with an identified competitor and often follows on from internal benchmarking when the organisation judges that it has gone as far as it can internally. This is obviously more complex and difficult, particularly in terms of persuading a competitor to provide commercially sensitive data.

(3) Functional
This is a comparison of functions/activities amongst a selected group of companies operating in the same industry. There are again clear difficulties in obtaining data from

what may be direct competitors. Provided data may be obtained, it is a highly effective improvement tool.

(4)   Generic

This benchmarks activities amongst a selected group of businesses in unrelated industries. For example, the health service may wish to benchmark its catering services functionally amongst a range of trusts or hospitals, but also generically amongst other organisations in different industries. This type of benchmarking may be less problematical in terms of data acquisition as it need not require co-operation of direct competitors.

To be successful, benchmarking requires:

❑   understanding of existing processes and a clear identification of those to be benchmarked
❑   a shared understanding in the organisation of what the objectives are in relation to overall organisational aims
❑   thoughtful choice of the internal units or external organisations to benchmark against having regard to compatibility/suitability of characteristics
❑   a robust method of measurement – i.e. clearly defined metrics/performance indicators
❑   an effective way of collecting the data
❑   an analytical approach to comparing differences and identifying the reasons for them. It should be noted that this may involve isolating of externalities that may be at work
❑   a realistic approach to improvement planning including the setting of objectives that are challenging but realistic
❑   careful and sensitive consideration of the organisational implications of the process including the need to ensure that all stakeholders are fully 'signed up' to the process and understand what it means.

This implies a process of:

❑   analysis – self appraisal
❑   analysis – existing processes
❑   planning
❑   administration
❑   data collection
❑   analysis – of data
❑   analysis – of processes to identify reasons for differences
❑   planning – improvement action plan
❑   implementing the action plan
❑   review and iterate.

The main danger is for organisations to attempt to benchmark too many things, and there is clearly a premium on selecting the correct key processes. These should be activities that are key to overall performance. Difficulties associated with data collection may arise, and it is imperative that the organisation is realistic and recognises that it is better to have limited but reliable data. It is also essential to carry out pilot tests to verify the methodology.

In a time of rapid change it is also important for the organisation to recognise that it may have shifting objectives, so that the shelf life of a specific benchmarking exercise may be quite limited and/or it may require constant overhaul.

Benchmarking does of course lead to the creation of league tables, and the use of these is the subject of much criticism and controversy. The major weakness with a league table approach is that it may give rise to a situation where we are not comparing like with like, a particular shortcoming of school performance league tables.

In order to be really effective benchmarking needs to be open and transparent, particularly in the public sector. If this objective is achieved it can provide excellent material upon which performance can be evaluated by all stakeholders, including customers, who can be provided with a meaningful measure of the extent to which improvements have been made.

Key performance indicators

A key performance indicator (KPI) is a measurement of the performance of an activity or parameter that is a critical success factor for an organisation, or perhaps an individual project. The widespread use of KPIs in the construction industry was driven by the 'Achieving Excellence in Construction' initiative launched in March 1999, which was itself prompted by both the Latham and Egan reports, in which parameters for continuous improvement of the industry's performance were clearly identified. The underlying thrust is one of obtaining better value for money including in the areas of refurbishment and maintenance.

The initiative set out a 'route map' with clear targets in relation to management, measurement, standardisation and integration. The key areas are identified in Table 9.1.

The 'Constructing Excellence' concept adopted the following mission:

To deliver individual, corporate and industry excellence in construction.[2]

To do this, it aimed to influence government by working with a number of stakeholder groups:

- ❏ DTI
- ❏ Strategic Forum for Construction
- ❏ central and local government
- ❏ ODPM
- ❏ private sector clients
- ❏ construction industry companies and organisations
- ❏ regional development authorities
- ❏ the research community
- ❏ trade and professional organisations,

Table 9.1 KPI Route Map.

Management
- ❏ Commitment and leadership
- ❏ Empowerment and skilling
- ❏ Consistent and skilled project management

Measurement
- ❏ Standard key performance indicators
- ❏ Post-project implementation reviews
- ❏ Client performance surveys

Standardisation
Key standard practices on:
- ❏ Procurement decisions on total value for money
- ❏ Use of risk and value management
- ❏ Output/performance specifications
- ❏ Whole life costing
- ❏ Robust change control
- ❏ Information technology and standardised document handling

Integration
- ❏ Teamwork and partnering
- ❏ Focus on design and build, PFI, design/build and maintain, prime contracting

(Source: DTI Constructing Excellence)

and acting as a catalyst for innovation, continuous improvement and the deployment of state-of-the-art technology.

To facilitate the process, a number of contributory organisations were founded together and comprehensive websites developed as a vehicle for the reporting, dissemination and analysis of information.

The Constructing Excellence team are grouped into four integrated programmes:

- ❏ innovation
- ❏ productivity
- ❏ best practice
- ❏ engagement with people, business and organisations.

Within this framework, it supports specific initiatives with particular sector groupings:

- ❏ the Housing Forum
- ❏ the Local Government Task Force
- ❏ the Infrastructure Task Force
- ❏ central government clients.

A major lead in implementing and driving improvement is the Movement for Innovation (M4I). The movement offers support to organisations wishing to use key performance indicators to benchmark performance. At present, wall charts and toolkits are available for:

- ❏ All Construction (Economic)
- ❏ Respect for People
- ❏ Environment
- ❏ Construction Consultants
- ❏ M and E Contractors
- ❏ Construction Products Industry
- ❏ Housing New Build
- ❏ Housing Repair and Maintenance and Refurbishment.

The indicators fall into three categories:

- ❏ headline indicators that provide a measure of the overall state of health of the company
- ❏ operational indicators that relate to specific activities within an organisation
- ❏ diagnostic indicators that provide intelligence help to explain why changes may have taken place in headline or operational indicators.

Table 9.2, extracted from a KPI report, gives a number of indicative examples of each of these.

Table 9.2 Hierarchy of KPIs.

Group	Indicator	Level
Time	Time for construction	Headline
	Time predictability – design	Headline
	Time predictability – construction	Headline
	Time predictability – design and construction	Operational
	Time predictability – construction (client change of instruction/order)	Diagnostic
Cost	Cost predictability – design	Headline
	Cost predictability – construction (client changes instruction/order)	Diagnostic
Quality	Defects	Headline
	Quality issues at availability for use	Operational
Client satisfaction	Client satisfaction – product	Headline
	Client satisfaction – service	Headline
Business performance	Profit	Headline
	Return on capital	Operational
	Repeat business	Diagnostic
Health and Safety	Reportable accidents	Headline
	Lost-time accidents	Operational

(Source: DTI Constructing Excellence)

As a further illustration of how the system operates, the example of client satisfaction is convenient. At present the Construction Clients' Forum, through the use of satisfaction surveys, records both headline indicators using the following scoring system:

10 = totally satisfied
5/6 = neutral
1 = totally dissatisfied

Given a client satisfaction score based on this information, the company is able to benchmark itself against the industry as a whole from data compiled by the group in the manner shown in figure 9.2.

In this example the company scores 8, and using the chart provided by the KPI group they can determine a benchmark score of 80%. This means that 20% of companies score higher on customer satisfaction.

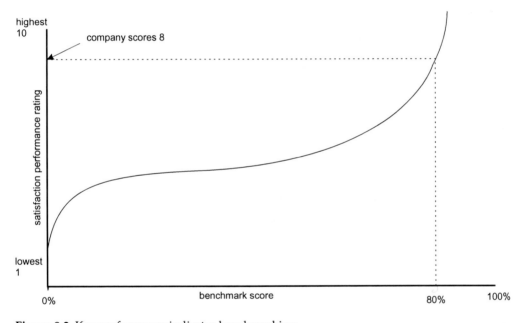

Figure 9.2 Key performance indicator benchmarking.

The balanced scorecard

Measuring public sector performance has always been fraught with difficulty and this has led ultimately to a drive towards performance management rather than performance measurement. NHS Estates have devised systems and guidance for trusts through metrics such as estate return data and patient environment assessments. The purpose has been to permit ready comparisons to be made in a form that is familiar to the customer, and to be able to evaluate compliance with management targets.

Because it is difficult for a large organisation to come to terms with what is in essence a multi-dimensional issue, the balanced scorecard concept has been identified as a methodology for addressing this, particularly in the context of an increasingly out-sourced (PPP or PFI) operation.

The balanced scorecard was initially conceived as a tool for the private profit-making organisation and originally developed by Kaplan and Norton.[3] The scorecard is an agreed set of measures, intrinsically linked to key strategies and priorities, that provide managers with a comprehensive view of performance.

In addition, because of its far-reaching effect, it is an important tool in understanding how an organisation works. The example given in figure 9.3 is taken from NHS Estates[4] and it shows that rather than focusing on a single aspect of success (or failure) there are a series of interconnected objectives and measures throughout the organisation. This provides a more holistic view by identifying patient outcomes, the enabling infra-structure required and resources required.

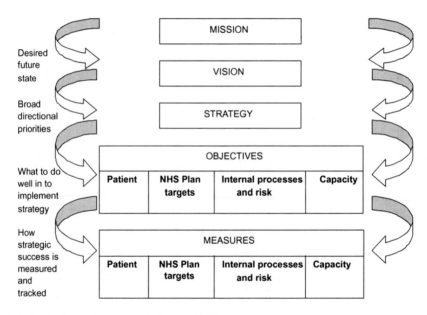

Figure 9.3 The balanced scorecard. *Source*: NHS

Organisation of maintenance departments

Whilst the relationship between a maintenance department and the rest of the corporate body can be extremely variable, it is possible to identify common elements or operations that will exist in the departments themselves. These include:

❑ identification of maintenance work, both planned and emergency, that can be called work input

- ❑ instructions for the work have to be transmitted to the work team
- ❑ execution, supervision, approval and valuation of the work
- ❑ authorisation and making payment
- ❑ a contractual framework
- ❑ an accounting context
- ❑ a feedback system, the sophistication of which is variable.

These can be represented in the maintenance system as illustrated in figure 9.4, which is based on Wiener's traditional view of an organisation as an adaptive system, entirely dependent on measurement and correction through information feedback.[5]

Maintenance organisations can be broadly encapsulated into two types of organisation:

- ❑ centralised
- ❑ decentralised.

There are significant differences in these models relating to operational aspects, but one of the most significant features relates to the size of organisation required at the centre. This will vary, not only with the policy adopted by the parent organisation, but also through a range of factors directly related to operational matters:

- ❑ the nature of the building stock
- ❑ volume, timing and diversity of the workload
- ❑ the complexity of the stock in technical terms
- ❑ geographical and topographical factors
- ❑ restrictions on the timing of work

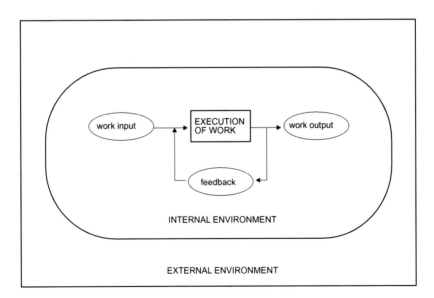

Figure 9.4 Maintenance as an adaptive system.

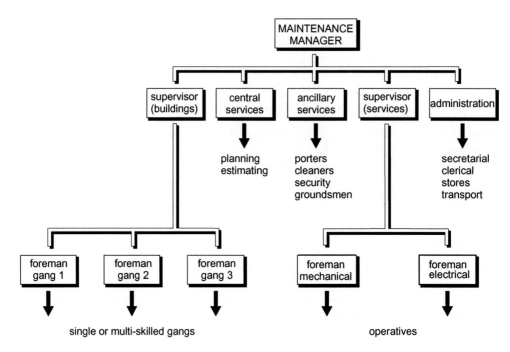

Figure 9.5 Centralised or functional maintenance organisation.

❏   whether work is to be executed by contract or direct labour
❏   the level of expertise of the workforce, and the extent to which non-operational
     tasks, such as routine inspections, are delegated to the people on the job
❏   how maintenance is defined in the organisation, such as the degree of involvement
     of the maintenance department in minor capital works.

The two examples shown in figures 9.5 and 9.6 allow for the possibility of work by
contract or direct labour, although at the detailed level many management functions
will be affected by the balance between these two.

Initiation of maintenance requests

The key document for initiating individual maintenance items is the work order, or the
job ticket, irrespective of whether this is issued to direct or contract labour. It must be
remembered that this document, as well as initiating work, may be an essential component of the reporting and recording mechanism. Such documents may be generated
manually, within an organisation's normal administrative processes, or through a help
desk facility set up as part of an integrated facilities management system. A great
number of variations exist, but in general they are designed following similar
principles.

A range of works orders can be used within an organisation to suit varying requirements, but this approach is difficult to justify. It is better to simplify the works order

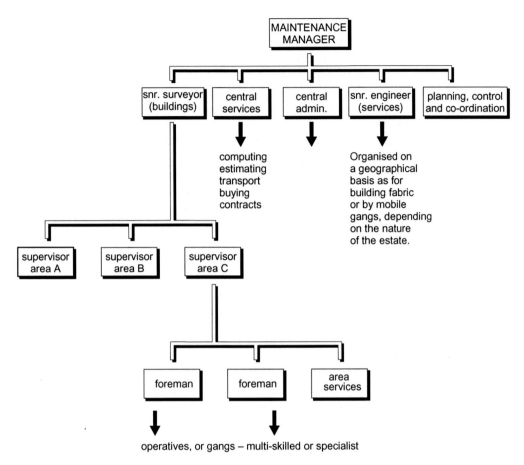

Figure 9.6 Decentralised or geographical maintenance organisation.

in order to assure uniformity, and recover feedback and control information by other means. For example, it should be possible to utilise the same basic works order for planned and unplanned work, and particularly in cases where maintenance gangs execute a mix of the two in any day this is a sensible approach. The following information should be included as a minimum.

(1) A job number and the date of issue.
(2) The address of the property or an appropriate location code.
(3) A work classification code, if one is in use.
(4) The location of the defect or service required.
(5) As accurate a description of the work as possible. The use of standard descriptions may be useful here, particularly if they can be coded in some way – perhaps linked to a classification code.
(6) Times and means of access, accompanied, perhaps, by a trigger code to indicate when permission to work is required.

(7)   For non-planned items, a priority coding will be necessary to allow teams to plan their work. In practice, this will increasingly be defined in a service level agreement. In principle, the following simple four-point approach is representative:

(i)   emergency work for which reasonable requests would receive at least essential attention within 24 hours

(ii)   urgent work that does not constitute a danger, but that should reasonably be completed in a week

(iii)   normal work, which will probably account for the major portion of the contingency allowance for non-planned work, and which may be included into a medium-term programme of work, perhaps on a three-monthly basis

(iv)   stand-by work, which is of low priority and can be held in reserve and fed into a planned programme at strategic intervals, in some cases, perhaps, to take up any slack in programmes.

In addition, a 'P' code can be entered against a planned item.

(8)   Estimated labour hours can be added to the job ticket, along with a space for the entry of the actual hours taken. This is facilitated if there is a library of standard jobs, coded, with a labour usage against it. This may not only be important from the point of view of control, but can also assist work planning. If a bonus system is operating there may also be a target time entered, which may not be the same as the estimated time. If contract labour is being used, then the information requirements on the order will reflect the contractual arrangement entered into. The entry of hours taken needs careful consideration, as there will inevitably be a proportion of non-productive time, for example due to travelling. This should be separated out for analysis and feedback purposes, and for comparison of actual response times against priority codes.

(9)   Estimated materials quantities may be entered; however, in all but simple cases this may be expecting too much of the system. A separate materials management system may be indicated.

Figure 9.7 illustrates the layout of a typical works order. Most electronically based maintenance management systems provide such an order, normally with the flexibility for customisation. The works order in an electronically based system will be generated by the system and permits a high level of analysis to take place.

Pre-execution functions

Once a job has been initiated there are certain management functions that need to take place before execution and afterwards. Before the works order is passed for execution any of the following may be necessary.

(1)   A check to ensure that the item has not been initiated by any other means – for example, an occupant request is not already in a planned programme.

(2)   A check to ensure that a request is valid, i.e. expenditure is justified, and there is no other conflict with work already initiated or planned for the future.

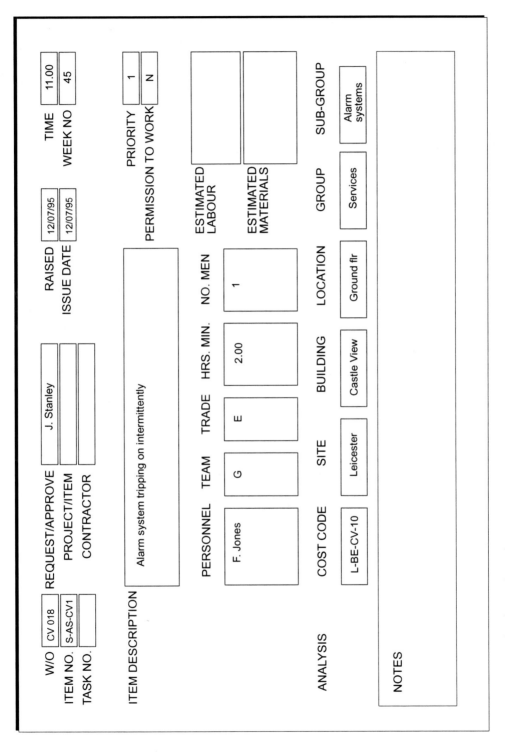

Figure 9.7 Example of a works order.

(3)   A decision made with respect to a priority rating. This is a difficult area, and it is convenient if this can be decided by the person receiving the request, so as to hasten the speed with which the job enters the system. The extent to which this is possible depends on organisational priorities.

(4)   A physical inspection of the defect or service required may be necessary for any of the following reasons:
- ❏   to estimate costs or resources required
- ❏   to arrange access
- ❏   to verify information for entry in the works order
- ❏   for diagnostic purposes.

The general principle adopted will tend to be to utilise exception principles, as one-off inspections add greatly to the cost of maintenance work. If work is of a low priority, however, such an inspection can be integrated into work execution.

The details of the process will obviously depend on the nature of the organisation and its building stock. The procedure outlined in figure 9.8 is a typical one, and again indicates the use of a database, which is a useful tool in checking for duplication.

Overlying all these requirements is the necessity to consider user requests as being very much the customer interface, and in terms of occupational efficiency and the performance of users it is unwise to underestimate physiological and psychological influences.

Post-execution functions

Following completion of the work, any of the following may take place:

(1)   An inspection of the work, to check that it satisfies requirements and to authorise payment. For contract labour the contract clauses will lay down certain requirements, as it is impractical to inspect every job in this way. In the case of contract labour, there has to be an implicit understanding that contractors will seek to provide a satisfactory service in order to retain the work. This is of particular importance in partnership arrangements where long-term agreements and relationships are guiding principles.

In the case of direct labour, it is a function of the maintenance department to establish a set of norms in terms of service expectations. It may have been decided, as a matter of policy, to inspect a range of jobs at random. It is desirable, however, that properly organised user feedback is introduced to provide sensible satisfaction intelligence.

(2)   Following a certified completion of the job, the property record will need to be updated.

(3)   The cost will be entered into an accounting system, either external to the maintenance department and/or internal, depending on the way in which costs are being recorded and analysed.

Figure 9.9 illustrates a typical procedure that might operate under an electronically driven database system.

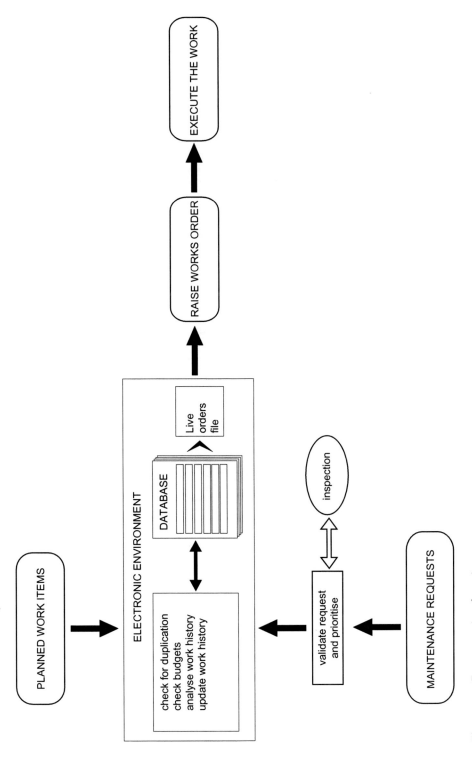

Figure 9.8 Pre-execution functions.

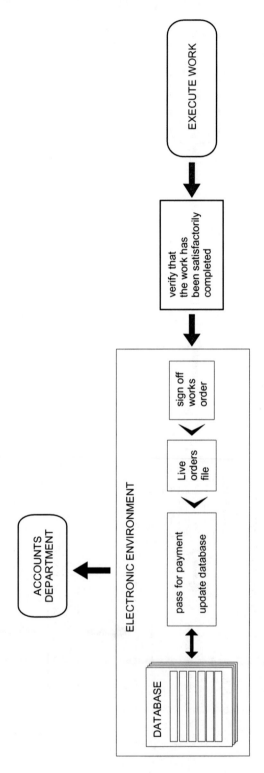

Figure 9.9 Post-execution functions.

Execution of work

There are a number of particular problems associated with the execution of maintenance work, which tend to revolve around the characteristic diversity of maintenance operations, the small-scale nature of individual items of work and the need for the attendance of a number of different trades. The problem of ensuring a proper sequence of trades, and balancing the relative numbers of each trade, is a productivity issue throughout the whole industry. As the size of an individual task reduces, there is a tendency for this problem to be exacerbated.

It may be assumed that planned maintenance programmes will be executed in a systematic way, and that reasonable planning of the sequence of operations and the balancing of the various trades is a realistic proposition, whether it be contract or direct labour.

There are several ways in which the labour force can be organised, and these are of equal relevance to both direct and contract labour.

(1)   A single group of operatives, consisting of one specific skill or specialism, may carry out a planned programme over a defined geographical area. An example of this might be a planned programme of flat roof replacement, carried out by an external contractor. This is called a *homogeneous contained* operation.

(2)   A problem occurs, however, when such an operation requires additional work, of a supportive nature, by another trade. A good example might be a planned programme of plumbing replacement work that requires a making-good and decorating operation following its completion. Under some circumstances it may be possible to attached a multi-skilled building operative to the team. This needs careful consideration to ensure a correct balance between the two types of work, and assumes the availability of such an operative. Some of the most obvious and commonly planned routine programmes may encounter this problem.

For example, redecoration may be accompanied by associated joinery or masonry work. If this has been identified prior to commencement some planning can take place; otherwise the act of redecoration will identify the need for work that was not foreseen. This is termed a *homogeneous uncontained* operation.

(3)   Other items of planned work will, by their very nature, involve a number of trades, and the planning problem will be more difficult, but still realistically possible. An example of this might be a window replacement programme that requires bricklayers, carpenters, glaziers and decorators. This is a *heterogeneous* operation. If each trade requires the same amount of time on each unit, then planning may be straightforward. Unfortunately this is rarely the case, so a balancing operation will be necessary.

(4)   There are other planned items that do not fit into any of these categories as they are not necessarily repetitive in nature. These will largely consist of items uncovered by planned inspections and low-priority maintenance requests, and will clearly need to be planned in rather a different way. These may be termed *semiplannable* items and, because of the organisational consequences, need separate consideration.

Ignoring these semi-plannable items for the time being, depending on the geographical dispersion of the estate, gangs can be organised on an area basis or by specialism.

(1) Homogeneous contained operations are probably better organised on the basis of a specialised gang operating on an estate-wide basis. Economies of scale suggest that subdivision is only realistic on very large estates.
(2) The organisation of homogeneous uncontained operations will depend on the amount of support work necessary. It has to be borne in mind here that the percentage of non-productive time involved in a making good operation may be very high, so that when this is only of a very minor nature there may be a case for regionalisation. Where the support work becomes more extensive, it may become realistic to have support operatives who travel rather further.
(3) The organisation of heterogeneous operations on a planned basis can be approached in two ways. Single trades may be rotated through the programme of work on an individual basis, or a balanced multi-trade gang may be set up. There are many advocates of the latter, on the basis that group relationships will form, and that benefits may flow from this. Ultimately, however, the viability of setting up such a gang depends rather on the nature of the operation.

Multi-trade gangs are more likely to be organised on an area basis, and to be responsible for more than one operation. Highly specialised gangs, on the other hand, will probably operate on a wider geographical basis. Much more troublesome is unplanned work, of which several types, presenting varying levels of organisational problem, can be identified.

(1) Items of work may be identified from a programme of planned inspection. Depending on the priority level, it may be possible to insert this work into a planned programme; that is, they can be termed semi-plannable.
(2) Items may be initiated by a user request which, depending on the level of priority, can become semi-plannable or, if very low priority, part of a fully planned longer-term programme. Many items from this source, though, will inevitably fall into the category of unplannable.
(3) Additional items uncovered by a normal planned programme present a large dilemma. They can be divided into two categories.
   ❑ Those whose rectification may be a pre-requisite for the completion of the primary maintenance operation, so that they require immediate treatment. These give rise to organisational problems, sometimes of a complex nature.
   ❑ Those treated like a user request, whose execution will depend on the priority attributed to them. It may, of course, be deemed appropriate to execute them immediately, purely on the grounds of efficiency, provided the planning of the primary item allows.

There are many items that will disrupt the careful organisation of work, and it can be seen that the distinction between planned and unplanned maintenance may become a little blurred at the execution level. Items that have been termed semi-plannable will not present an immediate problem, in that they will be inserted into a future pro-gramme in a controlled way. The main cost that has to be borne is the disruption to planning, and this emphasises the need for programmes to be dynamic and flexible.

Of greatest difficulty to organise are those items requiring immediate attention, and the two categories that have been identified above merit separate consideration.

Those immediate items uncovered during the execution of a planned operation can be exceedingly disruptive, and even with the most careful programme of inspection can never be completely eliminated. As well as the obvious repercussions for financial planning, they will inevitably disrupt the operation of a gang, often necessitate the calling on to the job of another trade, and sometimes force the temporary movement of the gang to another operation, possibly on another site.

The amount of disruption this all causes depends on the availability of another task appropriate to the gang concerned, and the response time of the additional labour involved. This highlights the benefits of a loose-fit style of management.

Urgent maintenance requests, not a prerequisite for the current operation, can be dealt with in a number of ways.

(1)   The setting up of multi-trade gangs that are purely executing planned items may not be realistic and is not, in any case, always the most effective way of organising the work. It is common practice for multi-trade gangs to receive a daily or weekly programme that is a mixture of planned and unplanned work. This provides one useful mechanism for the treatment of urgent repairs originating from users or in the field (figure 9.10). Under such a system, maintenance requests are logged and ordered according to priority rating, and allocated to each maintenance team according to their planned whereabouts and workload. Teams of this nature are likely to cover an area of an estate.

(2)   Maintenance teams may be very small, and in some cases just one or two multi-skilled persons can be of great value. There are advantages to be gained if these very small teams are based in a very localised area, as they can build up good working relations with users.

This is an approach advocated in housing management, where the local works office not only executes work, but receives maintenance requests and initiates work under rather looser central control.

It should be remembered, however, that some of the thinking behind this is sociological in nature, rather than based on maintenance efficiency considerations.

(3)   There will be a proportion of maintenance operatives who will solely execute emergency work, and these will almost certainly, except on very small estates, be responsible for a localised area.

The allocation of manpower solely to emergency work is a difficult process, and depends on being able to reasonably predict the volume of work and the response times required. There has been some experimentation with this type of work, particularly in housing maintenance, where small multi-skilled teams are employed, based on one estate with the objective of not only executing maintenance work but fostering good

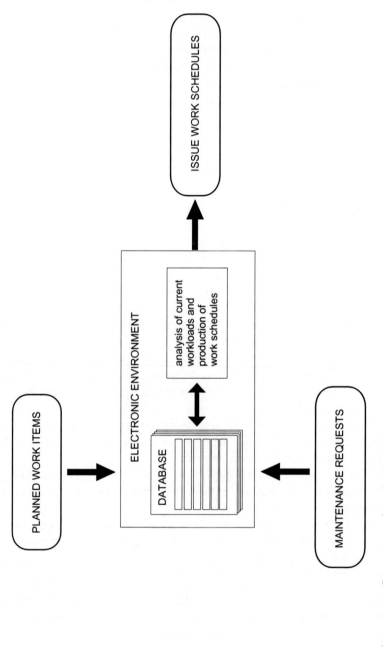

Figure 9.10 Generation of work schedules.

landlord–tenant relationships. This approach, it is hoped, will reduce vandalism, and hence reduce the number of maintenance requests. The economics of such small-scale operations are difficult to assess, and cannot be judged on raw maintenance efficiency criteria alone.

The maintenance operative

Maintenance operations are very labour intensive, and therefore the greatest potential for economising on the execution of the work lies in efficient use of labour. Studies have shown that for painting and decorating labour accounts for 85% of the prime cost, and for general repair work as much as 65%.[6]

The scope for improvement is largely to do with the efficient organisation of the work, as mechanisation has relatively little potential.

Somewhere in the region of 40% of the construction industry's workforce is employed in repair and maintenance. Amongst some trades the proportion involved in maintenance work is very high: for example, about 75% of all painters and decorators. The training of these operatives is largely along traditional lines, with very little specific to maintenance. A large volume of maintenance work is carried out within the traditional trade divisions, and many small contractors use the same labour for both new build and repair and maintenance.

It is quite normal for maintenance operatives to work in small gangs, made up of a number of trades, and there is evidence of a far greater blurring of the divisions between trades than occurs in new-build work. Furthermore, there are clear indications that the use of multi-skilled personnel can bestow large advantages in terms of flexibility of working. There are obstacles to these practices, and exceptions are more prevalent in Scotland and the North of England, and within local authorities.

Rigid trade demarcations clearly hinder the provision of an efficient and effective maintenance service, and concern about this on an industry-wide basis has been expressed in a series of reports dating back to the end of World War II. For example, as long ago as 1968 the Phelps Brown Report contained evidence commenting that an important part could be played in controlling the rising cost of maintenance by the encouragement of multi-craft skills.

Of major concern is the steady de-skilling of new construction, with the increasing tendency for buildings to become a series of assembly operations. This has contributed to the decline in training, in terms of volume, and also in the nature of the essential experience to which a trainee can become exposed.

As the level of skill required to execute work on existing buildings is generally higher than that required for the average new-build project, there is mounting concern that the supply of such skilled labour will diminish as the effects of reducing training programmes work their way into the system.

The labour force in maintenance tends to be older than that engaged in new build, and there are logical reasons behind this. The proportion of operatives over 40 years of age has been estimated as 58% for maintenance and only 32% for new construction.[7]

A newly qualified tradesman has high earning potential in his early twenties, and this potential is more likely to be realised in new construction. However, as he gets older, and perhaps undertakes additional family responsibilities, this earning potential is counterbalanced by the uncertain nature of his work, and he is likely to seek more security elsewhere in the industry. Maintenance work has tended to provide one of these more stable positions. This was, of course, particularly true in local authority DLOs, and although the situation has changed somewhat, maintenance output remains rather more stable than new construction.

It has been argued that the varied nature of maintenance work provides a higher level of personal satisfaction, particularly as there is a close relationship to the actual use of the building. There is a suggestion, therefore, that the maintenance operative is less motivated by purely financial reward, although it is difficult to separate these motivations from what many believe to be the disincentives of larger construction sites.

In proposing the need for a multi-skilled operative, it is necessary to be aware not only of practical difficulties, but also of the need to determine what basic skills are required. Maintenance involves such a wide range of activity that it is difficult to produce the perfect maintenance technician. The need for specialists will always exist, but many of these skills may be rather more contained. There is a strong case for arguing that the real provision of the type of skills required will come through experience and on-the-job training, rather than in a classroom. Individual craft courses can, however, do more to educate trainees in the broad nature of all construction work, so that their role can be seen in context.

Many of the demarcation problems that occur in maintenance are related to the need to execute support work. Making-good exercises are extremely common, and the skill level required for them is not necessarily high. The average carpenter is almost certainly quite capable of making good a small area of brickwork after fitting a new window.[8]

Productivity

There are a number of ways in which productivity can be measured, although it is a notoriously difficult subject on an industry-wide basis, as the interdependence of gangs and trades, and other characteristics of construction operations, always make it difficult to pin down responsibility. Maintenance work provides even more problems than new construction in this respect, and raw productivity data, no matter how accurate it may be, is by no means a firm indicator of the health or otherwise of a maintenance department.

In general, productivity depends on the work rate and the proportion of attendance time spent on productive work, and there are a number of measures that may give an indication of labour productivity.

(1)   Gross output per operative
This is given by the total cost of wages, materials and overheads, divided by the number of operatives employed. It gives a simple crude evaluation over the organisation as a whole, but is of little value for inter-trade comparisons.

The ratio is also vulnerable to overheads, in that an increase in overheads suggests an increase in productivity. The work output measure may also be affected by political considerations, or budgetary constraints. A cut in budget that restricts output reduces productivity as measured by this method.

(2)   Value of material used
This can be expressed as a total figure over a period, to permit comparisons of output over time, and/or in terms of material output per gang. For the purposes of comparing gang performance it is of little value, as the material usage of each gang varies owing to factors other than output. The figure can also be distorted by high wastage levels. Its main use is for evaluating the general level of output over time, in a similar fashion to the way brick stocks are used for the industry in general.

(3)   The rate of production of repetitive jobs
For repetitive jobs it may be possible to do simple comparisons, either in terms of the number of jobs executed in a given time or by an analysis of the times taken to complete jobs. There are obvious limitations to the appropriateness of this type of measure.

(4)   Comparisons with programmed or estimated hours
There are several variations on this theme, which have close links with work study techniques. The simplest approach is to divide the estimated time by the actual time taken. A more sophisticated variation is the development of a set of data that expresses the work content in terms of a number of standard hours. These are defined as the amount of work that can be performed in one hour by a standard skilled and motivated worker.

For example, the performance factor is equal to the standard hours of work produced divided by the actual hours of work expended. Other measures include cost per standard hour and gross cost of an hour's actual work.

In all these methods, there is a reliance on the derivation of either estimated times for executing work, or a standard hour, both of which present obvious difficulties. Within this is also some assumption that one job will always be like another, and whilst some element of repetition can reasonably be assumed, this is by no means as simple for maintenance work as would be expected. There will also always be items of work for which standard output targets cannot be derived, and their presence alone will distort productivity measurements.

There are other factors that can be evaluated. For example, one indicator of the effectiveness of a planned policy is to identify, for operatives or gangs, the number of hours spent directly on planned work, and compare this with the total hours spent. Interpretation of the results needs to be cautious, but this may be a useful management indicator, provided the full range of possible reasons for poor performance are investigated.

It is also possible to determine the amount of lost time incurred by a gang, and again this might indicate management shortcomings and problems with work allocation procedures as well as inefficient operative practices. If a proper reporting procedure has been instituted, it should be possible to classify lost time into one of the following causes:

- ❏ waiting for instructions
- ❏ waiting for materials
- ❏ waiting for access
- ❏ travelling time
- ❏ other.

The Audit Commission[9] advocate the determination of the percentage of productive time. They comment that it would be normal to expect this figure to be about 80%, but add that its measurement is subject to a great deal of variation, and this makes for considerable difficulty in comparing organisations.

None of these measures can be taken as particularly useful in absolute terms, but they all have some value in identifying trends and lend themselves to a benchmarking-style approach. The important thing is to always seek to discover the real underlying reasons for a poor performance in any respect.

Raw productivity in maintenance work will always tend to be lower than for new-build work because of a number of factors:

- ❏ the small-scale nature of the majority of individual jobs, leading to a high overall percentage of non-productive time
- ❏ the diversity of job content, which means that demarcation delays can be extensive
- ❏ the low level of repetition in unplanned work
- ❏ the wide dispersal of individual sites
- ❏ often difficult working positions
- ❏ the cost of obtaining access to areas for maintenance is a particular problem; scaffolding, for example, can cost many times more than the actual repair itself
- ❏ the ancillary work, such as making good, associated with a maintenance operation
- ❏ the sometimes high cost of obtaining materials to match existing work in older buildings.

Because of all these factors there is increased pressure on the efficiency of planning, the adoption of proper inspection routines to minimise disruption to planned programmes and the adoption of appropriate working methods. A number of Audit Commission reports have been frankly critical of the general quality of estate management in the public sector on these grounds.

Supervision

There are problems associated with the supervision of construction work in general, and this is exacerbated by the characteristics of maintenance work. There has been a tendency to criticise new construction sites for not providing sufficient back-up staff on a site, whereas in the maintenance sector, and particularly where direct labour is employed, it has been to comment unfavourably on the proportion of non-productive staff.

Within maintenance organisations the roles allotted to supervisors can vary greatly, but for simplicity it is useful to consider here the management who have direct responsibility for work execution and its control, including productivity.

The organisation models outlined earlier demonstrate the extreme approaches that can be taken but, in general, it can reasonably be assumed that the maintenance supervisor will be mobile, and that therefore a large amount of responsibility resides at the work face, in the hands of his operatives. His relationship with them is therefore of prime importance.

In many cases, the supervisor's role can also become a justifiably broader one, and encompass the whole range of attributes that help ensure an effective maintenance service.

References

(1) Audit Commission for Local Authorities in England and Wales (1989) *Building Maintenance Direct Labour Organisations – A Management Handbook*. HMSO, London.
(2) DTI (2004) *Constructing Excellence – A strategy for the future*. DTI Prospectus. www.dti.gov.uk
(3) Kaplan, R. & Norton, D. (1996) *The Balanced Scorecard*. Harvard Business Press, Boston, MA.
(4) NHS Estates (2003) *Guidelines for Implementing the Balanced Scorecard in NHS facilities and Estates*. NHS Estates, London.
(5) Wiener, N. (1948) *Cybernetics*. MIT, Boston, MA.
(6) Holmes, R. (1983) *Feedback on Housing Maintenance – BMCIS Occasional paper*. RICS, London.
(7) Department of the Environment, Welsh Office and Scottish Office (1993) *Housing and Construction Statistics 1983–1993*. HMSO, London.
(8) Audit Commission for Local Authorities in England and Wales (1989) *Building Maintenance Direct Labour Organisations – A Management Handbook*. HMSO, London.
(9) Audit Commission for Local Authorities in England and Wales (1989) *Building Maintenance Direct Labour Organisations – A Management Handbook*. HMSO, London.

Appendix 1
Statistics

In the day-to-day running of a building maintenance department a great deal of data will be generated. If this data is to be of any real value, the maintenance manager will need to know how to collect, organise, analyse, interpret and present it effectively. To do this, a basic knowledge of statistics is necessary, and to illustrate the point some basic principles are outlined in this appendix, using maintenance-related data.

Collection of data

Maintenance operations, by their very nature, generate large quantities of data, much of which will be recorded, i.e. it is collected. How much information is actually collected depends on management objectives.

Even when a comprehensive data set is available, it may be considered necessary to only use a sample of it for an analysis. The selection of this sample is an important consideration, as its size and nature will influence the results and conclusions drawn from its analysis. To obtain a meaningful result, the sample would have to be large enough to be significant, and chosen so as to be representative.

Presentation of data

Statistics is concerned with data of all kinds and, leaving aside the question of its analysis and interpretation, a primary requirement is to present it in the most suitable manner.

Tables are the simplest way of presenting information, but there are numerous approaches possible. Figure A1.1 shows statistics for local authority DLOs in the period 1970–1980. Another example, shown in figure A1.2, is taken from a study of the maintenance costs of CLASP buildings.

Tables of this nature can clearly become very complex, and simplified, rather more pictorial techniques can be used. For example, block diagrams or pie charts are useful ways of describing the maintenance costs of a building, and pictograms – such as that

	% of total public and private sector work				% of public sector work			
	all work	new work	repair and maintenance		new work			repair and maintenance
			housing	other	total	housing	other	other than housing
1970	10.5	3.5	17.7	37.1	6.8	6.4	6.9	54.0
1971	10.4	3.3	18.0	38.0	6.8	6.1	7.1	54.5
1972	10.4	3.2	18.2	37.7	6.8	5.6	7.4	52.8
1973	9.1	2.3	16.4	34.4	6.1	5.3	6.4	49.3
1974	9.2	2.5	16.4	33.9	5.4	5.1	5.6	50.6
1975	11.3	2.7	22.7	39.2	5.5	5.3	5.6	56.2
1976	11.8	3.1	20.8	39.0	6.1	5.7	6.3	56.8
1977	12.1	3.0	24.0	38.2	6.0	6.1	6.1	56.8
1978	11.4	2.7	21.3	35.7	6.2	5.9	6.3	54.4
1979	11.4	2.3	20.2	35.2	5.6	5.6	5.6	53.7
1980	11.1	2.3	19.7	32.6	5.6	5.8	5.5	49.5
1981	12.8	2.3	20.8	35.2	5.8	6.1	5.8	53.5
1982								
1983								
1984								

Figure A1.1 Tabulated data on DLO output based on housing and construction statistics.

year	age of building at end of year	structure	roofs	external finishes and claddings	external decoration	internal repairs	internal decoration	electrical services	plumbing services	heating services	external works	drainage	total maintenance
1962	6				603	18		46	11	72	20		770
1963	7			25		21		19	5	140	22		232
1964	8					30		28	7	168	209		442
1965	9		11	36		14		5	5	164	323	14	572
1966	10			5	217	62		22	19	35	119		479
1967	11		16	164		43	517	63	11	78	85		977
1968	12		11	10		30		9	14	62	82		218
1969	13		15	107		9			31	68	45		275
1970	14		4	94		74		12	25	151	101		461
TOTAL			57	441	820	301	517	204	128	938	1006	14	4426
average			6	49	91	33	57	23	14	104	112	2	492
percent			1.29	9.96	18.53	6.80	11.68	4.61	2.89	21.19	22.73	0.31	100.0

Figure A1.2 Study of maintenance costs for a school.

shown in figure A1.3, which gives the age profile of a housing association's housing stock – are also convenient for illustrative purposes.

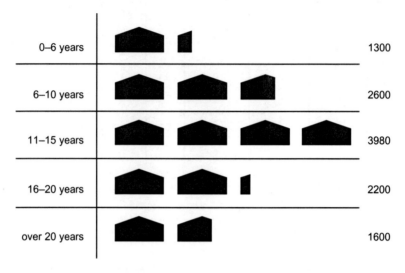

0–6 years	1300
6–10 years	2600
11–15 years	3980
16–20 years	2200
over 20 years	1600

Figure A1.3 Age composition of housing stock.

Frequency distributions

Much statistical data is concerned with the analysis of events or occurrences. For example, in maintenance operations, it may be necessary to study data on the frequency of the failure of components, or the frequency of call-outs for emergency repairs.

Figure A1.4 shows the results of studying the failure time of a component for a group of 50 houses over a 15-year period. It is assumed, in this example, that the component has only failed once in every unit. Table (a) is not very satisfactory, and a rearrangement as shown in (b) is more informative. Alternatively, the data may be grouped, as shown in (c), or presented graphically in any of the ways illustrated in figure A1.5.

The failure time that appears in this data set varies continuously, i.e. it is a continuous variable. The raw data is presented in a simplified form as, in practice, few if any of these failures occur after a precise number of years. The data only shows the years in which failure occurred.

Clearly a far more precise and complex set of data may be available, giving the precise time of failure, e.g. 3.2 years. Time is an example of a continuous variable but, for the purposes of representation, has been converted to a discrete variable.

Other variables may already exist in a discrete form, such as the number of emergency monthly call-outs received by a maintenance depot over a 24-day working month, illustrated in figure A1.6.

9	10	7	6	9	4	9	5	8	10
7	8	8	7	6	2	7	5	13	9
9	11	8	11	3	10	7	15	8	7
8	6	9	5	4	14	7	11	7	8
5	12	7	4	12	8	9	4	12	8

(a) Component failure years for 50 houses.

year to failure	number of houses
1	0
2	1
3	1
4	4
5	4
6	3
7	9
8	9
9	7
10	3
11	3
12	3
13	1
14	1
15	1

year to failure	number of houses in band
1–3	2
4–6	11
7–9	25
10–12	9
13–15	3

(b) Component failures
grouped by years.

(c) Component failures
grouped in year bands.

Figure A1.4 Analysis of component failure times.

When continuous variables are being converted to discrete ones, some decision has to be made concerning the range or precision to be used.

When presenting data, it is also important to be able to make simple comparisons. Figure A1.6 shows emergency calls for a maintenance depot in two separate months. A month-to-month analysis can be made by comparison of the actual call-out figures, or by their relative frequency, expressed in percentage terms. This can be represented graphically, by simple histograms or frequency polygons, as in figure A1.7.

Statistics over time

A time series analysis studies a single variable quantity at intervals over a period of time. There are a number of simple examples of this in Chapter 2, where construction

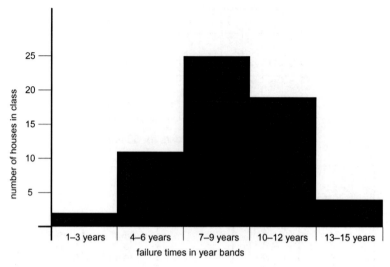

(a) Failure times as frequency histogram.

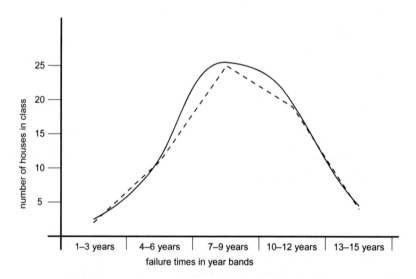

(b) Failure times as frequency polygon and frequency curve.

Figure A1.5 Analysis of component failure times.

MONTH 6		
day	no. calls	% total
1	3	1.7
2	5	2.8
3	7	3.9
4	4	2.2
5	9	5.1
6	6	3.4
7	4	2.2
8	9	5.1
9	5	2.8
10	7	3.9
11	9	5.1
12	8	4.5
13	5	2.8
14	4	2.2
15	7	3.9
16	6	3.4
17	9	5.1
18	8	4.5
19	10	5.6
20	9	5.1
21	11	6.2
22	11	6.2
23	12	6.7
24	10	5.6

MONTH 7		
day	no. calls	% total
1	12	5.3
2	9	3.9
3	11	4.9
4	10	4.4
5	9	3.9
6	11	4.9
7	10	4.4
8	9	3.9
9	8	3.5
10	9	3.9
11	9	3.9
12	10	4.4
13	8	3.5
14	9	3.9
15	6	2.6
16	8	3.5
17	9	3.9
18	10	4.4
19	9	3.9
20	10	4.4
21	11	4.9
22	10	4.4
23	11	4.9
24	10	4.4

Figure A1.6 Tabulation of two months' call-out figures.

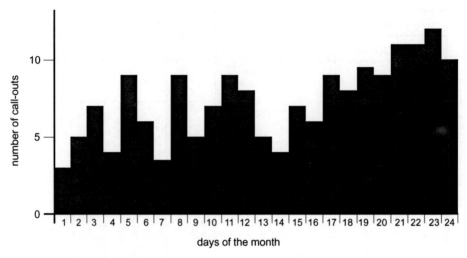

(a) Histogram of call-outs – month 6.

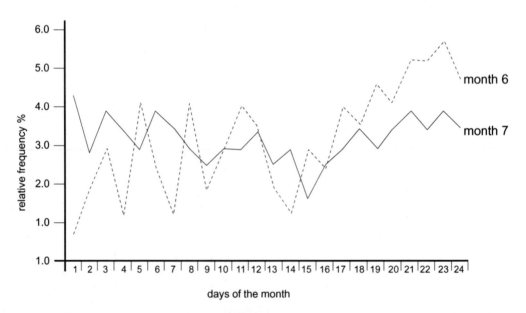

(b) Relative frequency polygon of call-outs.

Figure A1.7 Graphical representation of call-out figures.

output data is represented in a simple graphical format. In all these cases intervals of a year are involved, and although expenditure is a continuous variable, it is discretised into annual sums. It may also be useful to study the maintenance costs for a building in a similar way, or to analyse annual maintenance costs by discretising costs on a monthly basis. A further step may be to examine relative monthly costs (figure A1.8).

For carrying out comparisons, simple graphical representations are useful, but there is some value in the use of index numbers. These also have the additional merit of measuring the change in a variable with respect to a base point in time. Figure A1.9 compares the maintenance costs of two buildings over time in this manner.

	Expenditure	% total
January	650	5.1
February	826	6.5
March	745	5.8
April	950	7.5
May	916	7.2
June	1550	12.1
July	1675	13.1
August	1725	13.5
September	1440	11.3
October	950	7.5
November	721	5.7
December	603	4.7
TOTAL	£12751	100.0

Figure A1.8 Monthly breakdown of annual maintenance costs.

Cumulative representation

Statistical data, whether time-related or not, as well as being presented in an incremental way, can also be presented in a cumulative fashion. Figure A1.10 shows the monthly expenditure from figure A1.9 represented in this way, and illustrates how this form of presentation gives a visual indication of the rate of expenditure from the slope of the polygon.

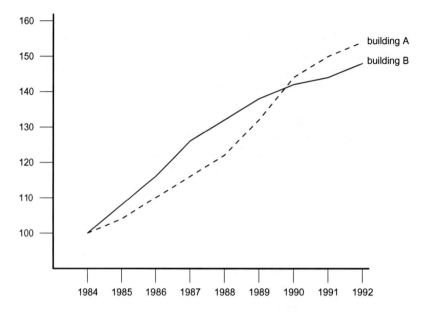

Figure A1.9 Indexed comparative annual maintenance costs at 1985 prices – 1984 base year = 100.

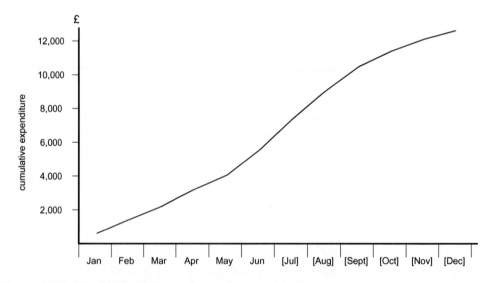

Figure A1.10 Monthly breakdown of annual maintenance costs.

The description of data

The presentation of data in a variety of ways allows the formulation of a qualitative view, from which inferences of a general nature may be drawn. To permit quantitative judgements, it is necessary to be able to characterise a data set by more objective means. If a frequency distribution, with grouping of data representing component lives, is available, an arithmetical mean can be determined as shown in figure A1.11.

Whilst the mean is the simplest, and most widely understood, measure it can often be a very misleading one, as extreme values can have a disproportionate effect. This is easily demonstrated in the following example. In carrying out a cost-in-use exercise, the following component lives in years are obtained from maintenance data, and the problem is one of deciding which should be used for evaluative purposes:

$$3, 3, 2, 9, 3, 9, 3, 2, 3, \ 2, 10, 3, 3, 2, 3$$

The average of these is slightly over 4, whereas 12 out of the 15 figures are less than this. In this simple case it may be decided to discard the figures that are obviously causing the distortion. However, for the purpose of analysing a data set objectively, there are other measures that can be used.

Life to failure	Frequency F	Midpoint X	F × X
1–3	12	2	24
4–6	15	5	75
7–9	35	8	280
10–12	20	11	220
13–15	10	14	140

ΣF = 92	Σ FX = 739

Average life = ΣF / Σ FX

= 739/87

= 8.03 YEARS

Note that this calculation assumes
an even distribution within each class.

Figure A1.11, Calculation of mean component life.

The median

If the above figures are arranged in order of magnitude:

$$2, 2, 2, 2, 3, 3, 3, \mathbf{3}, 3, 3, 3, 3, 9, 9, 10$$

the median is the middle value in the series, i.e. 3.

Where larger amounts of data are involved, and perhaps grouped, the median can be found from the frequency polygon. Figure A1.12 shows how the median divides the area under the polygon in half, and it is possible to carry out a similar exercise to produce other divisions, and hence determine quartiles, or a variety of other percentiles. The cumulative frequency polygon may be easy to evaluate for these purposes.

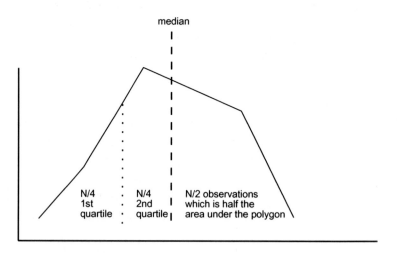

(a) Frequency polygon.

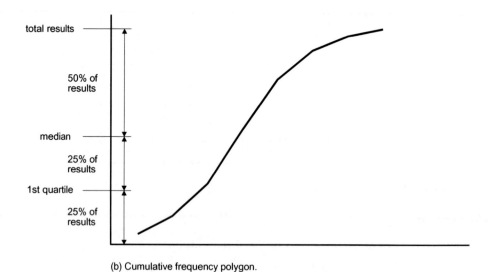

(b) Cumulative frequency polygon.

Figure A1.12 Median and quartiles.

The mode

The mode is another simple measure that may be more useful than an arithmetic average, and is the value that occurs most frequently in a series. In the simple example given above, the mode is 3. For a larger data set it can very easily be determined from a frequency polygon (figure A1.13).

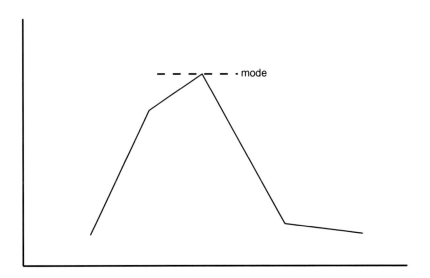

Figure A1.13 The mode from a frequency polygon.

Scatter, spread and dispersion

In analysing and comparing the number of daily call-outs for two different areas of an estate, the frequency polygons shown in A1.14 are produced. Analysis of the two sets of figures indicates that the average daily call-out number is roughly the same, yet their characteristic distributions are clearly different.

A measure of variability is needed to describe accurately the characteristics of a distribution. One way of doing this might be might be to calculate the average deviation from the mean of each value in a data set. However, as figure A1.15 demonstrates, this creates a problem. It is a mathematical certainty that, if signs are taken into account, the positive and negative deviations will always cancel out. This can be overcome by ignoring the minus signs, which produces what is called a mean deviation. Of far greater benefit, however, is the standard deviation.

Standard deviation

The standard deviation can be calculated as shown in A1.15. Frequency curves are commonly used to describe graphically the characteristics of a data set, and the

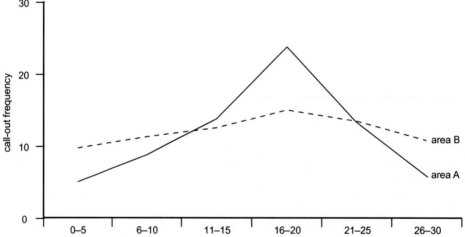

Area A			
Call-outs	Frequency F	Midpoint X	F × X
1–5	5	3	15
6–10	9	8	72
11–15	14	13	182
16–20	24	18	432
21–25	14	23	322
26–30	6	28	168

| ΣF = 72 | ΣFX = 1191 |

Average call-outs = ΣFX / ΣF

= 1191/72

= 16.54

Note that this calculation assumes an even distribution within each class

Area B			
Call-outs	Frequency F	Midpoint X	F × X
1–5	10	3	30
6–10	11	8	88
11–15	13	13	169
16–20	15	18	270
21–25	13	23	299
26–30	11	28	308

| ΣF = 73 | ΣFX = 1164 |

Average call-outs = ΣFX / ΣF

= 1164/73

= 15.94

Note that this calculation assumes an even distribution within each class

Figure A1.14 Comparative daily call-out figures for two areas over a three-month period.

standard deviation. Natural phenomena, and events of chance, usually produce frequency curves that are bell-shaped, as shown in figure A1.16. In this example, the two curves represent data sets equal in size, with the same arithmetic mean. The standard deviation, therefore, gives a means of distinguishing between the two.

house number	maint. costs x	deviation from mean mean − x	deviation squared (mean − x)²
1	198	15.6	243.36
2	186	27.6	761.76
3	209	4.6	21.16
4	206	7.6	57.76
5	211	2.6	6.76
6	196	17.6	309.76
7	230	-16.4	268.96
8	200	13.6	184.96
9	225	-11.4	129.96
10	218	- 4.4	19.36
11	240	-26.4	696.96
12	245	-31.4	985.96
13	200	13.6	184.96
14	212	1.6	2.56
15	248	-34.4	1183.36
16	228	-14.4	207.36
17	237	-23.4	547.56
18	170	43.6	1900.96
19	205	8.6	73.96
20	208	5.6	31.36
total	= £4272	Σ(mean − x) = 0	Σ(mean − x)² = 7611.44

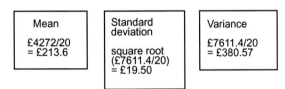

Mean	Standard deviation	Variance
£4272/20 = £213.6	square root (£7611.4/20) = £19.50	£7611.4/20 = £380.57

Figure A1.15 Analysis of maintenance costs of a sample of 20 houses.

Coefficient of variation

If two data sets have greatly different means, then the standard deviation is insufficient to distinguish them. The coefficient of variation expresses the standard deviation as a percentage of the mean.

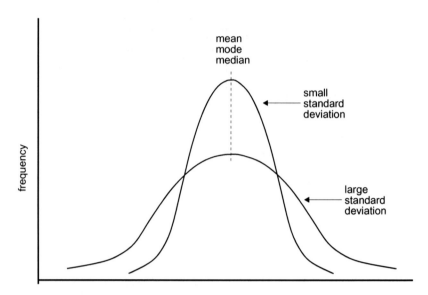

Figure A1.16 Comparative standard deviations.

Skewness

It is quite possible to have two sets of data that possess the same mean and the same standard deviation, but have quite different distributions about the mean. The distinguishing factor between the two sets of data shown in figure A1.17 is the degree of symmetry, or skewness. Skewed curves are often characteristic of economic or social observations.

Another frequency curve is the so-called bathtub frequency curve, which is common when considering the incidence of failure of building components through the life of a building.

The use of frequency distributions

The mathematical properties of frequency distributions can be described by an equation, in terms of a number of mathematical variables and constants including the following:

❏ the frequency of a given variable
❏ the total frequency of the distribution
❏ the standard deviation
❏ the mean of the distribution.

For a normal distribution, using tables derived from such equations, it is possible to determine the proportion of cases between two vertical lines on the curve, i.e. the percentage of the total sample that has a value within a given range.

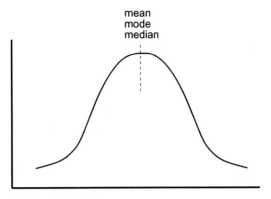

(a) Normal distribution.

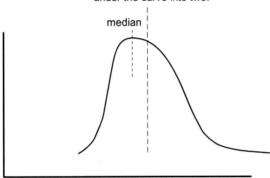

(b) Positively skewed distribution.

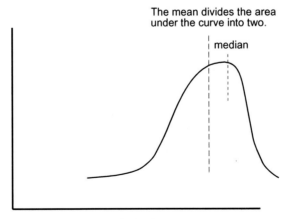

(c) Negatively skewed distribution.

Figure A1.17 Comparative distributions.

The principle is indicated in figure A1.18(a), with reference to what is called the standard normal distribution, which has a mean of zero and a standard deviation of one. For the general case, therefore, these percentages are in terms of multiples of the standard deviation (figure A1.18 (b)).

Considerable manipulation is possible by using the tables, and this is best illustrated by an example. Figure A1.19 represents data on the failure profile of strip lights in a factory, and in instituting a planned replacement policy it is necessary to consider a range of replacement cycles. This involves a management decision with respect to an acceptable chance of a failure between replacement cycles.

In the example shown, this is carried out using the normal distribution. The figures are derived from mathematical tables, found in a number of good statistical texts, where a fuller exposition of the mathematical principles can also be found.

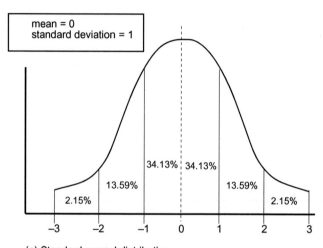

mean = 0
standard deviation = 1

34.13%   34.13%

13.59%   13.59%

2.15%   2.15%

−3   −2   −1   0   1   2   3

(a) Standard normal distribution.

99.74% of all cases lie between ∓ 3 standard deviations.

Choice of a value equal to the mean − 1.64 standard deviations means that only 5% of values will fall below this figure.

Choice of a value equal to the mean + 2.33 standard deviations means that only 1% of values will lie above this figure.

(b) General cases.

Figure A1.18 Analysis of the normal distribution.

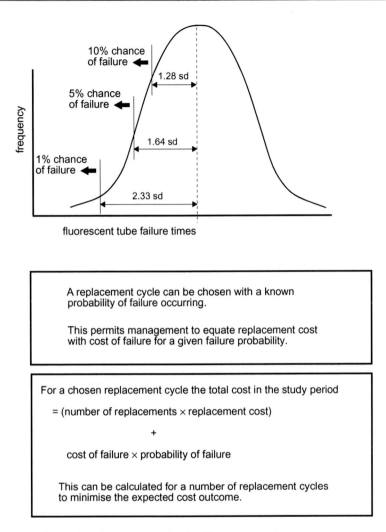

Figure A1.19 Analysis of replacement cycles for fluorescent tubes.

Similar exercises can be carried out on different distributions, but with a little more complexity. There are also a number of statistics computer packages that can be used, so an intimate knowledge of theory is not necessary to make use of these techniques.

Probability and expected outcomes

Following the above exercise, an evaluation of alternative strategies can be made, taking into account failure probabilities as shown in A1.19. It should be noted that a more rigorous analysis needs to take account of the time value of money. This introduces the notion of probability, along with related techniques used for decision-making, such as decision trees.

Correlation techniques

Everyday experience throws up numerous instances of the manner in which different phenomena are associated in some way. For example, in Chapter 2 the relationship between maintenance output and a range of economic indicators was examined, and reference made to correlation. In its simplest form, this branch of statistics deals with two variables, and seeks to establish the relationship between them.

A local authority knows full well that there is a relationship between the age of its schools and annual maintenance costs, but needs to establish this in more objective terms. Figure A1.20 illustrates the results of the authority plotting annual maintenance cost against age for its school stock in a particular area, and this is termed a scatter diagram. Assuming, rather loosely, that this relationship can be represented by a straight line, it is possible, using what are termed linear regression techniques, to determine the equation of the best straight line that fits this data, thus allowing extrapolation of maintenance costs into the future.

It is, of course, very debatable whether a straight line is appropriate, and it is possible to force a straight line on to any set of data. It is important to realise that this is only the best fit to the data under review. In A1.21 (a) there is an almost perfect fit whereas in (b) there is virtually none.

Such exercises, therefore, need to be qualified by other means. Before attempting the above exercise, it is wise to establish whether there is a numerical connection between the variables, and whether or not it is a linear one.

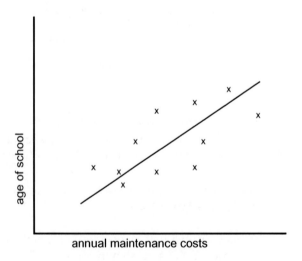

Figure A1.20 Scatter diagram for school maintenance costs.

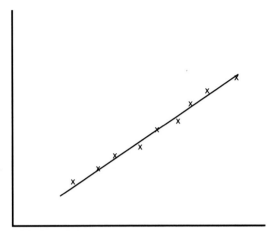

(a) Close fit to straight line.

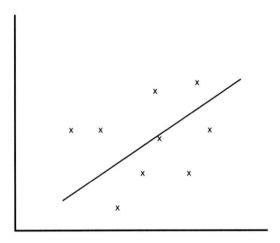

(b) Poor fit to straight line.

Figure A1.21 Scatter diagram fit to straight line.

The coefficient of correlation provides the measure of the degree of linear association present. It must be stressed that a poor correlation in this respect does not eliminate the possibility of a strong non-linear relationship, as shown in figure A1.22.

In some cases, normal direct plots may not show sensible linear correlation, but it should be borne in mind that alternatives are available such as:

❏   a plot of one variable against the square of another
❏   a plot of one variable against the log of another.

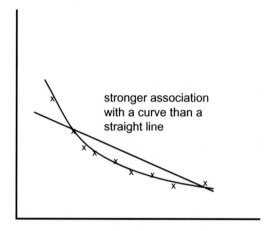

Figure A1.22 Curved versus straight line association.

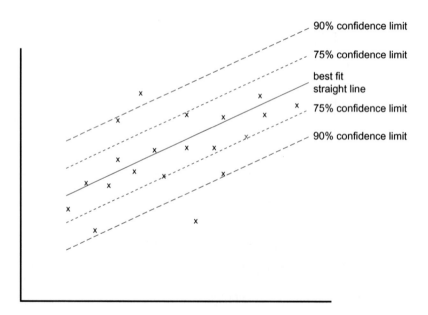

Figure A1.23 Confidence limits.

Such plots may produce a straight line even though a non-linear relationship exists.

Where a strong correlation is achieved, indicating a linear relationship, there will still be a degree of scatter. The outcome can be qualified, in such cases, by the statement of confidence limits as illustrated in A1.23.

Time series analyses

Indexing

Indexing is a commonly used technique for measuring trends, particularly costs, over a period of time. For example, the retail price index and the number of indices used for measuring construction costs are generally well known. The principle is demonstrated in A1.24, and the data could then be plotted on a graph. However, some care is needed when there are a number of contributory costs to be consolidated into one index.

Maintenance costs are made up of materials, labour and plant input, although the latter is often ignored. Indices are available, from a number of sources, for the individual inputs, but before they can be incorporated into a total maintenance cost index their relative contribution has to be determined. This produces what is called a weighted index, and is the method used by BMI, and referred to in Chapter 2. A simple example of such an exercise is illustrated in A1.25.

Trends with time

If a variable, such as material costs, is plotted over time, the true relationship may be masked somewhat by other variables. This means that the figures used may need to be adjusted to take account of such phenomena. For example, it might be expected that construction output would vary on a seasonal basis so that, when studying activity over a longer term, figures have to be seasonally adjusted.

In considering trends, the effects of random unforeseen events, such as damage caused by a storm of great magnitude that would only be expected to occur, perhaps, once every

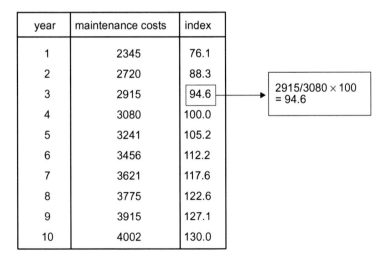

year	maintenance costs	index
1	2345	76.1
2	2720	88.3
3	2915	94.6
4	3080	100.0
5	3241	105.2
6	3456	112.2
7	3621	117.6
8	3775	122.6
9	3915	127.1
10	4002	130.0

$2915/3080 \times 100 = 94.6$

Figure A1.24 Indexing of maintenance costs.

year	labour cost index	material cost index	weighted overall cost index
1	91	88	90.0
2	94	92	93.3
3	97	97	97.0
4	100	100	100.0
5	103	105	103.7
6	108	110	108.7
7	115	118	116.1
8	126	125	125.7
9	130	129	129.7
10	135	133	134.3

Labour : Materials
relative inputs 65 : 35

(91×0.65)
$+$
(88×0.35)

$= 90$

Figure A1.25 Determination of weighted index costs.

100 years, need to be considered. If it happens to occur in the period under study, then clearly a distorted picture might emerge. There will also be cyclical variations that effect trends, if they are not in phase.

The use of moving averages does much to eradicate some of these variables, and there are a number of other statistical techniques that are widely used.

Appendix 2
Methods of Financial Appraisal

The value of money over time

There are a number of financial appraisal techniques, ranging from the simple to the sophisticated, that can be of use as an aid to decision-making in many areas of building design and evaluation, as well as during the management of the building in use. The simplest technique, the straightforward payback approach, suffers from a failure to take into account the time value of money. Offered £1000 today, or the same amount in a year's time, it will be beneficial to take the money now, on the basis that it can be invested, and therefore will be worth more in the future.

At an interest rate of 10%, £1000 invested today will be worth £1100 in a year's time. Viewed conversely it can be said that £1100 in a year's time has a present value of £1000.

The sum of £1000 to be received in a year's time has a present value of:

$$£1000 \times \frac{100}{110} = £909.10$$

In mathematical terms, the present value (PV) of £1 after n years with an interest rate of:

$$i = 1/(1 + i)^n$$

In practice, tables of discount factors are available, to enable future cash flows to be discounted to a present value. Figure A2.1 shows an extract from such a table.

Thus, if the cost of replacing a component in ten years' time is £1500 then the present values of replacement are:

$$PV = £1500 \times 0.5584 = £837.60 @ 6\% \text{ discount rate}$$

$$PV = £1500 \times 0.4224 = £633.60 @ 9\% \text{ discount rate}$$

Many costs are recurring annual ones, and whilst each year could be discounted separately, there is also a formula, and corresponding table, that facilitates the determination of the present value of a regular series of cash flows over a period of years. An extract of a table is shown in figure A2.2.

Present value of £1

no. of years	interest	
	6%	9%
5	0.7473	0.6499
10	0.5584	0.4224
20	0.3118	0.1784
30	0.0303	0.0057

Figure A2.1 Present value (PV) of £1 in the future.

Present value of £1 per annum

no. of years	interest	
	6%	9%
5	4.212	3.890
10	7.360	6.418
20	11.470	9.129
30	13.765	10.274
60	16.161	11.048

Figure A2.2 Present value (PV) of £1 per annum in the future.

Thus, if there is an annual cleaning cost, over a presumed 60-year building life, of £1500 per annum, the respective present values are:

$$PV = £1500 \times 16.161 = £24\,241.50 \ @ \ 6\% \text{ discount rate}$$

$$PV = £1500 \times 11.048 = £16\,572.00 \ @ \ 9\% \text{ discount rate}$$

There are a number of variations on the same theme, all based around the same mathematical principles.

Net present values

The net present value technique (NPV) involves discounting all future cash flows to a common base year. Suppose that a design team is faced with the following scenario.

There are two alternative ways of providing the cladding to a building. Alternative A involves an initial outlay of £100000 with predicted annual maintenance costs of £2000. Alternative B has an initial outlay of £120000 with predicted annual mainte-nance costs of £500.

The cash flow, and discounted equivalents, at a rate of 9%, are shown in figure A2.3.

Whilst, in simple cash flow terms, option A appears to be the most expensive option, when the figures are discounted it has the lower present value. This can be put another way. An investment now of £122 096 will pay the total cost of alternative A over its design life, but an equivalent investment now of £125 524 will be required to pay for option B.

Alternative A

cash flow	year(s)	discount factor	present value
£100 000	1	1	£100 000
£ 2 000 (×60 yrs)	1–60	11.048	£ 22 096
£220 000			£122 096

Alternative B

cash flow	year(s)	discount factor	present value
£120 000	1	1	£120 000
£ 500 (×60 yrs)	1–60	11.048	£ 5 524
£150 000			£125 524

Figure A2.3 Comparison of alternatives using present values.

Annual equivalent

Another approach that can be taken, based on the same principles, is the so-called annual equivalent method. In this technique, the cash flows throughout the life of an asset are converted into an equivalent annual cost. In terms of evaluation, it will rank alternatives in exactly the same order as the net present value approach, but will present the figures in a more meaningful way for the building owner.

To obtain the annual equivalent, the NPV is divided by the PV of £1 per annum for the appropriate number of years. For example, to convert the PV for alternative A above to an annual equivalent, £122096 divided by 11.048 gives £11051. In other words, an expenditure of £11051 per year over 60 years has a present value of £122096.

The usefulness of the technique for life cycle costing is shown in figure A2.4, where a series of cash flows over time, that are not recurring in phase, can be converted to an equivalent annual expenditure.

Softwood windows require a capital outlay of £4500 and renewal every 15 years at a cost of £5000 at today's prices, including an allowance for removing the old ones.

Redecoration will be required every five years at a cost of £400. The economic life of the building is taken to be 60 years.

The annual equivalent cost is determined below.

cash flow	year(s)	discount factor	present value
£4500	0	1.0000	£4500
£ 400	5	0.6499	£ 260
£ 400	10	0.4224	£ 269
£5000	15	0.2745	£1373
£ 400	20	0.1784	£ 71
£ 400	25	0.1160	£ 46
£5000	30	0.0754	£ 377
£ 400	35	0.0490	£ 20
£ 400	40	0.0318	£ 13
£5000	45	0.0207	£ 104
£ 400	50	0.0134	£ 5
£ 400	55	0.0087	£ 3

total present value £6941

60 year £1 per annum discount factor 11.048

Annual Equivalent = £6941/11.048

annual equivalent £ 628

Figure A2.4 Determination of annual equivalent over 60-year life.

Internal rate of return

Another derivative technique is the determination of a so-called internal rate of return. The object of the exercise, in this case, is to determine the interest rate that will produce, when all future cash flows, positive and negative, are taken into account, a net present value of zero; that is, when discounted costs equate to discounted benefits.

Determination of the internal rate of return is most simply carried out by determining the NPV of a set of cash flows at various discount rates and plotting them onto a graph, from which the IRR can be found.

Sinking funds

When allowance has to be made to meet a known future capital expenditure, one of the more prudent ways of doing this is to set up a sinking fund. This involves setting aside a regular sum of money that, when invested, will accumulate sufficiently to meet that future commitment. A good example is the requirement for housing associations to allow for major renewal programmes.

The requirement in this case is to determine the amount of money that needs to be set aside annually, at a given discount rate, to amount to the capital requirement in a number of years' time.

An industrial organisation predicts that it needs to carry out a major refurbishment to a production unit in five years' time, which will cost £100000.

Determine the amount it needs to invest in a sinking fund to achieve this, assuming a discount rate of 8%.

Discount tables provide a uniform series sinking fund factor.

For five years at 8%, this is 0.17045

The required amount per year is therefore:

$$£100000 \times 0.17045 = £17045.$$

In other words, the setting aside of £17045 each year for the next five years will accumulate to £100000 at an interest rate of 8%.

Index

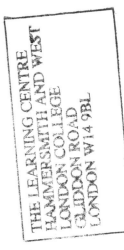

Lightning Source UK Ltd.
Milton Keynes UK
UKOW040610031111

181389UK00002B/21/P